Autodesk Revit for Architecture Certified User Exam Preparation

Daniel John Stine

SDC
PUBLICATIONS

SDC Publications
P.O. Box 1334
Mission, KS 66222
913-262-2664
www.SDCpublications.com
Publisher: Stephen Schroff

Examination Copies
Books received as examination copies are for review purposes only and may not be made available for student use. Resale of examination copies is prohibited.

Electronic Files
Any electronic files associated with this book are licensed to the original user only. These files may not be transferred to any other party.

Trademarks
Autodesk Revit is a registered trademark of Autodesk, Inc. All other trademarks are trademarks of their respective holders.

The author and publisher of this book have used their best efforts in preparing this book. These efforts include the development, research and testing of the material presented. The author and publisher shall not be liable in any event for incidental or consequential damages with, or arising out of, the furnishing, performance, or use of the material.

ISBN-13: 978-1-63057-408-6
ISBN-10: 1-63057-408-2

Printed and bound in the United States of America.

Foreword

The intent of this book is to provide the reader with a **study guide** and electronic **practice exam** (download) to prepare for the Autodesk Certified User, Revit for architecture exam. You will find an overview of the exam process, the user interface and the four main topics: Elements/Families, Modeling, Views, and Documentation. At the end of the book, there is a sample multiple-choice **practice test** to self-assess your readiness for the exam.

While this study guide cannot claim to cover every possible question that may arise in the exam, it does help to firm up your basic knowledge to positively deal with most questions… thus, providing more time to reflect on the more difficult questions.

> **Errata:**
> Please check the publisher's website from time to time for any errors or typos found in this book after it went to the printer. Simply browse to www.SDCpublications.com, and then navigate to the page for this book. Click the **View/Submit errata** link in the upper right corner of the page. If you find an error, please submit it so we can correct it in the next edition.
>
> You may contact the publisher with comments or suggestions at service@SDCpublications.com.

Trial and Student Software:

This book is based on Autodesk *Revit 2022*. A **20-day** trial may be downloaded from Autodesk's website. Additionally, qualifying students may download the free **3-year** student version of the software from students.autodesk.com. Both are fully functional versions of the software. The provided practice exam, as well as the official exam, require the use of Revit and the provided Revit project files to successfully answer most questions.

About the Author

Daniel John Stine AIA, CSI, CDT, is a registered architect with over twenty years of experience in the field of architecture. He is the Director of Design Technology at the top ranked architecture firm Lake|Flato in San Antonio, Texas. Dan implemented BIM-based lighting analysis using ElumTools, early energy modeling using Autodesk Insight, virtual reality (VR) using the HTC Vive/Oculus Rift along with Fuzor & Enscape, Augmented Reality (AR) using the Microsoft HoloLens 2, and the Electrical Productivity Pack for Revit (sold by ATG under the CTC Software brand). Dell, the world-renowned computer company, created a video highlighting his implementation of VR at LHB.

Dan has presented internationally on BIM in the USA, Canada, Scotland, Ireland, Denmark, Slovenia, Australia and Singapore; Autodesk University, RTC/BILT, Midwest University, AUGI CAD Camp, NVIDIA GPU Technology Conference, Lightfair, and AIA-MN Convention. By invitation, he spent a week at Autodesk's largest R&D facility in Shanghai, China, to beta test and brainstorm new Revit features.

Engaged professionally, Dan is a member of the American Institute of Architects (AIA), Construction Specifications Institute (CSI), and Autodesk Developer Network (ADN), Chair of the Illuminating Engineering Society (IES) BIM Standards committee and is a Construction Document Technician (issued by CSI). He has presented live webinars for Enscape, ElumTools, ArchVision, Revizto and

NVIDIA. Dan has also been a guest on several podcasts, including Business of Architecture, BIM Thoughts, and Simply Complex.

Committed to furthering the design profession, Dan teaches graduate architecture students at North Dakota State University (NDSU) and has lectured for interior design programs at NDSU, Northern Iowa State, UTSA, and University of Minnesota, as well as Dunwoody's new School of Architecture in Minneapolis. As an adjunct instructor, Dan previously taught AutoCAD and Revit for twelve years at Lake Superior College.

Dan writes about design on his blog, BIM Chapters, and in his textbooks published by SDC Publications:

- *Residential Design Using Autodesk Revit 2022*
- *Commercial Design Using Autodesk Revit 2022*
- *Design Integration Using Autodesk Revit 2022 (Architecture, Structure and MEP)*
- *Interior Design Using Autodesk Revit 2022 (with co-author Aaron Hansen)*
- *Autodesk Revit 2021 Architectural Command Reference (with co-author Jeff Hanson)*
- *Residential Design Using AutoCAD 2022*
- *Commercial Design Using AutoCAD 2013*
- *Chapters in Architectural Drawing (with co-author Steven H. McNeill, AIA, LEED AP)*
- *Interior Design using Hand Sketching, SketchUp and Photoshop (also with Steven H. McNeill)*
- *SketchUp 8 for Interior Designers; Just the Basics*

Certification Study Guides

The author of this book has also prepared the following certification study guides students can use to achieve successful results when taking certification exams which can help made the difference when trying to secure that dream job!

- *Microsoft Office Specialist Excel Associate 365/2019 Exam Preparation*
- *Microsoft Office Specialist Word Associate 365/2019 Introduction & Exam Preparation*

You may contact the publisher with comments or suggestions at service@SDCpublications.com.

BIM Chapters:

Daniel Stine's blog: http://bimchapters.blogspot.com/

Social Media:

Students can use social media, such as Twitter and LinkedIn, to start developing professional contacts and knowledge. Follow the author on social media for new articles, tips and errata updates. Also, consider following the design firms and associations (AIA, CSI, etc.) in your area; this could give you an edge in an interview!

 Twitter
@DanStine_MN

 LinkedIn
https://www.linkedin.com/in/danstinemn

Many thanks go out to Stephen Schroff and SDC Publications for making this book possible!

Table of Contents

Downloads:

- **Practice Test Software**
  See inside-front cover for download instructions and your unique access code

- **Practice Exam Software**
  See inside-front cover for download instructions and your unique access code

1.0 Introduction

Overview

In the competitive world in which we live it is important to stand out to potential employers and prove your capabilities. One way to do this is by passing one of the Autodesk Certification Exams. A candidate who passes an exam has credentials from the makers of the software that you know how to use their software. This can help employers narrow down the list of potential interviewees when looking for candidates.

When the exam is successfully passed a certificate can be printed and displayed at your desk or included with your resume. You also have access to an Autodesk logo for use on business cards or on flyers promoting your work; two examples are shown below as well.

The exams are <u>not</u> based on a specific release of Revit. Rather, they are based on fundamental Revit concepts which typically do not change from version to version. Thus, it does not matter which version of Revit you have installed on your personal computer or in the lab at school.

Important Things to Know

Here are a few big picture things you should keep in mind:

- **Practice Exam**
    - The practice exam, that comes with this book, is taken on **your own computer**
    - You need to have **Revit installed** and ready to use during the practice exam
    - You must download the practice exam software from SDC Publications
        - See inside-front cover of this book for access instructions
    - **Required Revit files** for the practice test
        - Files downloaded with practice exam software
        - Locate files before starting practice test
    - Note which questions you got wrong, and study those topics

- **Autodesk Certified User (ACU) Exam**
    - Purchase the **exam voucher** ahead of time
        - If you buy it the day of the test, or at the test center, there may be an issue with the voucher showing up in your account
        - Note: some testing locations charge an extra proctoring fee.
    - Make a **reservation** at a test center; walk-ins are not allowed
    - A computer is provided at the test center
    - Have your Certiport **username** and **password** memorized (or written down)
    - If you fail, note which sections you had trouble with and study those topics
    - You must wait 24 hours before retaking the exam

Benefits

There are a variety of reasons and benefits to getting certified. They range from a school/employer requirement to professional development and resume building. Whatever the reason, there is really no downside to this effort if you are, or hope to be, working in the architecture or construction industry.

Here are the benefits listed on Autodesk's website:

- Earn an industry-recognized credential that helps prove your skill level and can get you hired.
- Develop your skills with sample projects and exercises that emphasize real-world applications.
- Accelerate your professional development and help enhance your credibility and career success.
- Validate your skills and join an elite team of Autodesk Certified professionals.
- Display your Autodesk Certified certificate, use the Autodesk Certified logo, highlight your achievement and get noticed by listing your name in the Autodesk Certified Professionals database.

Certificate

When the ACU exam is successfully passed, a certificate signed by Autodesk's CEO is issued with your name on it. This can be framed and displayed at your desk, copied and included with a resume (if appropriate) or brought to an interview (not the framed version, just a copy!).

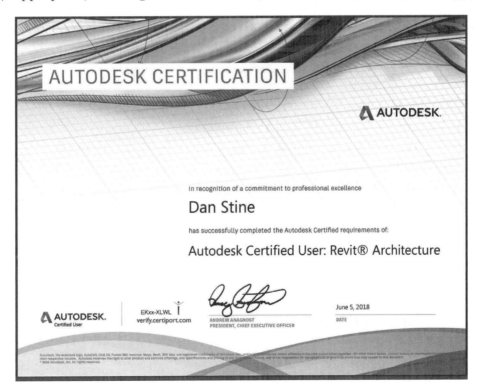

A high-quality **print** of your official certificate may be purchased via certiport.com in the My Transcript section (log in using the same username and password for the exam).

Badging

In addition to a certificate, a badge is issued. Badging is a digital web-enabled version of your credential by **Acclaim**, which can be helpful to potential employers. This is a quick proof that you know how to use the Revit features covered by the ACU exam.

Certified User *versus* Certified Professional

While this study guide focuses solely on the **Revit for Architecture - Autodesk Certified User** (ACU) exam, it is helpful to know about the other options for future consideration. There are five different exam options, plus a practice test. These are all paid options, not free, but when considering the value outlined previously, it may be worth it. See the links at the end of this section to learn more about costs.

Autodesk Certified User (ACU) exam for Revit.
Autodesk Certified User (ACU) certification is an excellent way for students with about 40 hours of real-world Autodesk software experience to validate their software skills.

- ACU: **30 questions** which must be answered in **50 minutes.**
  Passing: 70% which means at least 21 questions must be answered correctly

Autodesk Professional Certification (ACP) exam with Four discipline-specific options.
Autodesk Certified Professional (ACP) is the logical credential for advanced and professional Autodesk software users who possess at least 400 hours of real-world Autodesk software experience.

- Revit for Architecture: In addition to the features common to all disciplines of Revit, this exam is focused on the architectural tools within Revit, such as Walls, Doors, Windows, Toposurface, Building Pads, Rooms, etc. There are 35 questions which must be answered in 120 minutes.

- Revit for Electrical Building Systems: This exam is focused on the electrical tools within Revit, such as electrical panels, devices, equipment, MEP systems, Spaces, etc. There are 35 questions which must be answered in 120 minutes.

- Revit for Mechanical Building Systems: This exam is focused on the mechanical tools within Revit, such as ducts, pipe, air terminals, equipment, MEP systems, Spaces, etc. There are 35 questions which must be answered in 120 minutes.

- Revit for Structures: This exam is focused on the structural tools within Revit, such as beams, beam systems, columns, analytical model, loads, etc. There are 35 questions which must be answered in 120 minutes.

Both the ACU and ACP exams utilize a combination of multiple choice and live in-the-application style questions.

Autodesk Certified User (ACU)

Autodesk Certified User (ACU) certification is an excellent way for students with about 40 hours of real-world Autodesk software experience to validate their software skills.

Learn more»

Autodesk Certified Professional (ACP)

Autodesk Certified Professional (ACP) is the logical credential for advanced and professional Autodesk software uses who possess at least 400 hours of real-world Autodesk software experience.

Learn more»

Exam Topics and Objectives

The ACU exam covers four main topics related to Content, Modeling, Views, and Documentation. The outline below lists the specific topics one needs to be familiar with to pass the test. The remainder of this book expounds upon each of these items.

Creating & Modifying Components

- Create and modify grids
- Create and modify levels
- Create and modify walls
- Load and modify doors
- Load and modify Windows
- Tag components by category
- Load and modify components

Modeling and Modifying Elements

- Create a roof and modify roofs
- Create and modify stairs
- Create and modify ramps
- Create and modify railings
- Create and modify floors
- Modify elements using align, offset, mirror, and split tools
- Modify elements using move, copy, rotate, trim, and extend tools

Managing Views

- Change the view scale
- Change the detail level of a view
- Manage visibility/graphics overrides for model categories
- Manage view range
- Duplicate views
- Create section views
- Create elevation views
- Create 3D views and renderings

Managing Documentation

- Create and modify text
- Create and modify dimensions
- Create and modify a sheet
- Place plan views on a sheet
- Create and modify schedules

Exam Releases (including languages)

The **Certiport** website lists which languages and units of measure the exam & practice tests are available in as partially shown in the image below. For the full list, follow this link:

https://certiport.pearsonvue.com/Educator-resources/Exam-details/Exam-releases

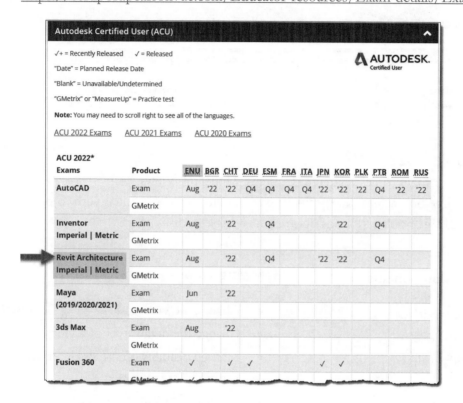

Exam releases

Certified Training Centers

There are several places to take the exam. Many academic institutions administer the exam directly to their students. Several Autodesk Resellers also offer the exam. And there are formal testing facilities which offer a full range of similar exams, from Yoga Instructor certification to Senior Pharmacy Technician certification.

To find the nearest testing center, start here: http://portal.certiport.com/Locator

Unfortunately, there may not be a test center in your city. For example, the closest testing facility for the author of this study guide is 150 miles away. In this case, you will have to plan a day to travel to the testing center to take the ACU exam. In this case it is much more important to have made an appointment, purchased the voucher ahead of time and associated with your Certiport account… and of course studied the material well, so you do not have to retake it.

Locating a training center

From the **Certiport** ACU frequently Asked Questions (FAQ) online page:

> "Educators and students can take the exams at a public Certiport Authorized Testing Center or become a center themselves.

> If schools or districts want to run exams onsite, they can easily become a testing center and run the exams seamlessly in class. Institutions can sign up to be centers on the Certiport site."

Practice Exam

A **practice exam** is included with this book which can be downloaded from the publisher's website using the **access code** found on the inside-front cover. This is a good way to check your skills prior to taking the official exam, as the intent is to offer similar types of questions in roughly the same format as the formal ACU exam. This practice exam is taken at home, work or school, on your own computer. You must have Revit installed to successfully answer the in-application questions.

This is a test drive for the exam process:

- Understanding the test software
- How to mark and return to questions
- Exam question format
- Live in-application steps
- How the results are presented at the exam conclusion

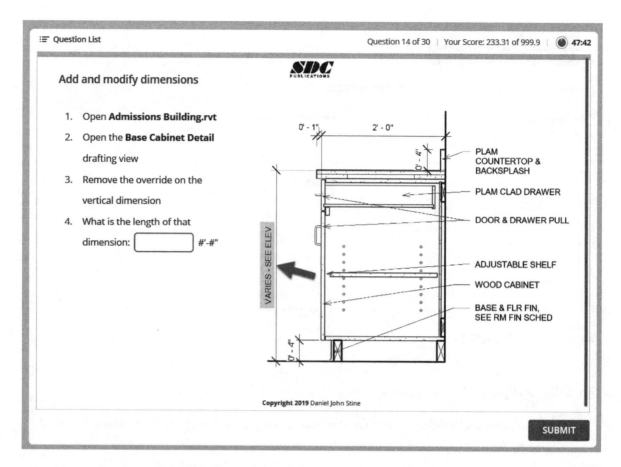

Add and modify dimensions

1. Open **Admissions Building.rvt**
2. Open the **Base Cabinet Detail** drafting view
3. Remove the override on the vertical dimension
4. What is the length of that dimension: [   ] #'-#"

0' - 1"   2' - 0"

0' - 4"

PLAM COUNTERTOP & BACKSPLASH

PLAM CLAD DRAWER

DOOR & DRAWER PULL

ADJUSTABLE SHELF

WOOD CABINET

BASE & FLR FIN, SEE RM FIN SCHED

VARIES - SEE ELEV

0' - 4"

Copyright **2019** Daniel John Stine

SUBMIT

Sample question from included practice exam

If you enter your name and email address at the beginning of the practice exam, you will receive an email with your result. If you are required to take the Practice exam for a class, this results email can then be forwarded to your instructor. Alternatively, you could also enter your instructor email address. Whether you enter your name and email address or not, neither the author nor SDC Publications captures any data related to this practice exam.

Having taken the practice test can remove some anxiety one may have going into an exam that may positively impact your career search.

This book also offers a **practice test**, several questions where Revit use is not required, and provides the answers to self-assess your readiness for the formal practice test and exam.

See chapter 10 for more details on the Practice Exam software provided with this book.

Exam Preparation

Before taking the Autodesk Certified User (ACU) exam, you can prepare by working through **this study guide**, taking the **practice test** and then the **practice exam**. You may also want to drive to the test location a day or so before the exam to make sure you know where it is and what the parking options are to ensure you are on time the day of the exam.

During the Exam

During the exam, be sure to manage your time. Quickly go through the test and answer the questions that are easy to you, skipping the ones you are not immediately sure of. The exam software allows you to view a list of questions you have not answered or have marked. Once you have answered all the easy questions you can then go back and think through those which remain. Do not exit the exam until you are completely finished, as you will not be able to re-enter the exam after that point.

Exam Results

Once the exam is finished you will immediately see your score. You must earn 700 points out of 1000 to pass (70%). If you failed, you should note the objective areas where you missed questions and study those areas more before taking the test again – see image below. Be sure to print your score report and take it with you to study – it is also possible to log into your Certiport account later and print it from home.

SECTION ANALYSIS	
Creating and Modifying Components	100%
Modeling and Modifying Elements	89%
Managing Views	100%
Managing Documentation	100%

FINAL SCORE	
Required Score	700
Your Score	957

OUTCOME	
Pass	✓

Retaking the Exam

If the exam is failed, don't worry as you can take it again – as soon as 24 hours later. If you have any doubt about your ability to easily pass the exam, consider purchasing a voucher that includes a reduced cost "retake" option.

In the event that you do not pass the exam, and you have purchased the retake option, a retake code will be emailed to you. You may re-take the exam after waiting 24 hours from the time your initial exam was first started. Retake vouchers must be used within 60 days of the failed exam.

Here is the currently posted retake policy for the certification exam:

- If a candidate does not achieve a passing score on any Autodesk Certified Professional exam the first time, the candidate must wait one day (24 hours) before retaking the exam.
- If a candidate does not achieve a passing score the second time, the candidate must wait five days (120 hours) before retaking the exam a third time.
- A five-day waiting period will be imposed for each subsequent exam retake.
- There is no annual limit on the number of attempts on the same exam.
- If a candidate achieves a passing score on an ACP exam, the candidate may take it again.
- This policy applies to both voucher-based and site-license-based Certiport Centers.
- All original vouchers must be used prior to their expiration dates, without exception.
- Retake vouchers must be used within 60 days of the initial failed exam. If a retake was purchased, the retake voucher is sent by email after a failed exam.
- Test results found to be in violation of this retake policy will result in the candidate not being awarded the attempted credential, regardless of score.

Resources

For more information visit these sites:

- Certiport: http://www.certiport.com/autodesk
- Autodesk Certification Site: www.autodesk.com/certification
- Autodesk Authorized Training Center (ATC): www.autodesk.com/atc
- Acclaim (Credly): https://www.youracclaim.com/

Certiport User Registration

Here are the steps to create a Certiport account, which is required to take the exam.

Start here: https://www.certiport.com/Portal/Pages/Registration.aspx

Follow the steps outlined on the site. Once complete, you will be prompted to register your account with a certification program. **Important:** be sure to select the Autodesk option in this step, and not Microsoft, Adobe, etc. This can be done later in your **My Profile** section (see image below).

Register your account with a certification program

Notes:

2.0 Revit Fundamentals

What is Autodesk Revit?

Autodesk Revit (Architecture, Structure and MEP) is the world's first fully parametric building design software. This revolutionary software, for the first time, truly takes architectural computer aided design beyond simply being a high-tech pencil. Revit is a product of Autodesk, makers of AutoCAD, Civil 3D, Inventor, 3DS Max, Maya and many other popular design programs.

Revit can be thought of as the foundation of a larger process called **Building Information Modeling** (BIM). The BIM process revolves around a virtual, information rich 3D model. In this model all the major building elements are represented and contain information such as manufacturer, model, cost, phase and much more. Once a model has been developed in Revit, third-party add-ins and applications can be used to further leverage the data. Some examples are Facilities Management, Analysis (Energy, Structural, Lighting), Construction Sequencing, Cost Estimating, Code Compliance and much more!

Revit can be an invaluable tool to designers when leveraged to its full potential. The iterative design process can be accomplished using special Revit features such as *Phasing* and *Design Options*. Material selections can be developed and attached to various elements in the model, where one simple change adjusts the wood from oak to maple throughout the project. The power of schedules may be used to determine quantities and document various parameters contained within content (this is the "I" in BIM, which stands for Information). Finally, the three-dimensional nature of a Revit-based model allows the designer to present compelling still images and animations. These graphics help to more clearly communicate the design intent to clients and other interested parties. This book will cover many of these tools and techniques to **assist** in the creative process.

What is a parametric building modeler?

Revit is a program designed from the ground up using state-of-the-art technology. The term parametric describes a process by which an element is modified and an adjacent element(s) is automatically modified to maintain a previously established relationship. For example, if a wall is moved, perpendicular walls will grow, or shrink, in length to remain attached to the related wall. Additionally, elements attached to the wall will move, such as wall cabinets, doors, windows, air grilles, etc.

Revit stands for **Rev**ise **In**stantly; a change made in one view is automatically updated in all other views and schedules. For example, if you move a door in an interior elevation view, the floor plan will automatically update. Or, if you delete a door, it will be deleted from all other views and schedules.

A major goal of Revit is to eliminate much of the repetitive and mundane tasks traditionally associated with CAD programs allowing more time for design, coordination and visualization. For example: all sheet numbers, elevation tags and reference bubbles are updated automatically when changed anywhere in the project. Therefore, it is difficult to find a mis-referenced detail tag.

The best way to understand how a parametric model works is to describe the Revit project file. A single Revit file contains your entire building project. Even though you mostly draw in 2D views, you are actually drawing in 3D. In fact, the entire building project is a 3D model. From this 3D model you can generate 2D elevations, 2D sections and perspective views. Therefore, when you delete a door in an elevation view you are actually deleting the door from the 3D model from which all 2D views are generated and automatically updated.

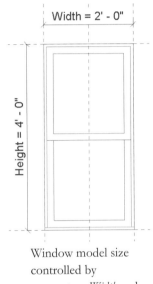

Another way in which Revit is a parametric building modeler is that parameters can be used to control the size and shape of geometry. For example, a window model can have two parameters set up which control the size of the window. Thus, from a window's properties it is possible to control the size of the window without using any of the drawing modify tools such as Scale or Move. Furthermore, the parameter settings (i.e., width and height in this example) can be saved within the window model (called a Family). You could have the 2′ x 4′ settings saved as "Type A" and the 2′ x 6′ as "Type B." Each saved list of values is called a Type within the Family. Thus, this one double-hung window Family could represent an unlimited number of window sizes! You will learn more about this later in the book.

Window model size controlled by parameters *Width* and

Example of window family rendered using Enscape, a real-time rendering add-in for Revit

Trial Software

If you do not have access to Revit to practice for the ACU exam, you can download a trial version. Autodesk offers a free 30-day trial of Revit. To learn more, follow this link: https://www.autodesk.com/products/revit/free-trial

Below are the minimum system requirements to run Revit:

OPERATING SYSTEM [1]	**Microsoft® Windows® 7 SP1 64-bit:** Enterprise, Ultimate, Professional, or Home Premium
	Microsoft Windows 8.1 64-bit: Enterprise, Pro, or Windows 8.1
	Microsoft Windows 10 Anniversary Update 64-bit (version 1607 or higher): Enterprise, or Pro
CPU TYPE	Single- or Multi-Core Intel® Pentium®, Xeon®, or i-Series processor or AMD® equivalent with SSE2 technology. Highest affordable CPU speed rating recommended. Autodesk Revit software products will use multiple cores for many tasks, using up to 16 cores for near-photorealistic rendering operations.
MEMORY	4 GB RAM • Usually sufficient for a typical editing session for a single model up to approximately 100 MB on disk. This estimate is based on internal testing and customer reports. Individual models will vary in their use of computer resources and performance characteristics. • Models created in previous versions of Revit software products may require more available memory for the one-time upgrade process.
VIDEO DISPLAY RESOLUTIONS	**Minimum:** 1280 x 1024 with true color

	Maximum: Ultra-High (4k) Definition Monitor
VIDEO ADAPTER	**Basic Graphics:** Display adapter capable of 24-bit color **Advanced Graphics:** DirectX® 11 capable graphics card with Shader Model 3
DISK SPACE	5 GB free disk space
MEDIA	Download or installation from DVD9 or USB key
POINTING DEVICE	MS-Mouse or 3Dconnexion® compliant device
BROWSER	Microsoft® Internet Explorer® 7.0 (or later)
CONNECTIVITY	Internet connection for license registration and prerequisite component download

Student Software

Students can download a free 3-year version of Revit at www.students.autodesk.com. Be sure to use your school email address. When using the educational version and logged in to Autodesk A360, using your school email address, you have access to unlimited cloud credits for rendering.

> The student version of the software is intended for learning purposes only; it is against the licensing agreement to use a student version for any professional work. If found using student licenses to do professional work, you may be subject to legal action from Autodesk.

The system requirements for the student version are the same as the trial version listed above.

3D model of lunch room created in Interior Design using Autodesk Revit 2019

Why use Revit?

Many people ask the question, why use Revit versus other programs? The answer can certainly vary depending on the situation and particular needs of an individual or organization.

Generally speaking, this is why most companies use Revit:

- Many designers and drafters are using Revit to streamline repetitive drafting tasks and focus more on designing and detailing a project.
- Revit is a very progressive program and offers many features for designing buildings. Revit is constantly being developed and Autodesk provides incremental upgrades and patches on a regular basis.
- Revit was designed specifically for architectural design and includes features like:
    - Photo Realistic Rendering
    - Phasing (different design over time)
    - Design Options (different designs within the same time period)
    - Live Schedules
        - Quantity Schedules
        - Material Takeoff Schedules
        - Sheet Index Schedule
    - Cloud Rendering via *Autodesk 360*
    - Analysis via *Autodesk 360*
        - Conceptual Energy Analysis
        - Daylighting Analysis
        - Structural Analysis

File Types and their extensions:

Revit has four primary types of files that you will work with as a Revit user. Each file type, as with any Microsoft Windows based program, has a specific three letter file name extension; that is, after the name of the file on your hard drive you will see a period and three letters:

.RVT	Revit project files; the file most used
	Backup files .RVT**.0001**, .RVT**.0002**, etc.
.RFA	Revit family file; loadable content for your project
.RTE	Revit template; a project starter file with office standards preset
.RFT	Revit family template; a family starter file with parameters

A few basic Revit concepts:

The following is meant to be a brief overview of the basic organization of Revit as a software application. You should not get too hung up on these concepts and terms as they will make more sense as you work through the tutorials in this book. This discussion is simply laying the groundwork so you have a general frame of reference on how Revit works.

The Revit platform has three fundamental types of elements:

- Model Elements
- Datum Elements
- View-Specific Elements

Model Elements

Think of *Model Elements* as things you can put your hands on once the building has been constructed. They are typically 3D but can sometimes be 2D. There are two types of *Model Elements*:

- **Host Elements (aka System Family)** Walls, floors, slabs, roofs, ceilings. These are items that can only exist within a project—they cannot be loaded from a file.

- **Model Components** (Stairs, Doors, Furniture, Beams, Columns, Pipes, Ducts, Light Fixtures, Model Lines) – Options vary depending on the "flavor" of Revit.

  - Some *Model Components* require a host before they can be placed within a project. For example, a window can only be placed in a host, which could be a wall, roof or floor depending on how the element was created. If the host is deleted, all hosted or dependent elements are automatically deleted.

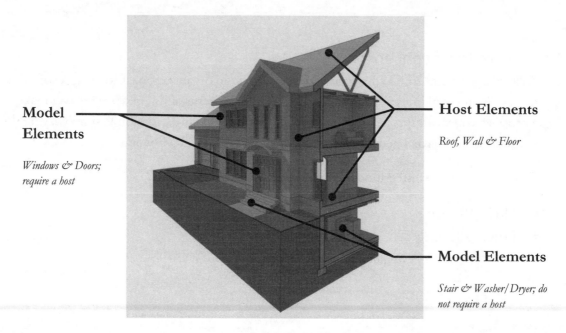

Model Elements

Windows & Doors; require a host

Host Elements

Roof, Wall & Floor

Model Elements

Stair & Washer/Dryer; do not require a host

Datum Elements

Datum Elements are reference planes within the building that graphically and parametrically define the location of various elements within the model. These features are available in all "flavors" of Revit. These are the three types of *Datum Elements*:

- **Grids**
  - ❖ Typically laid out in a plan view to locate structural elements such as columns and beams, as well as walls. Grids show up in plan, elevation and section views. Moving a grid in one view moves it in all other views as it is the same element. (See the next page for an example of a grid in plan view.)

- **Levels**
  - ❖ Used to define vertical relationships, mainly surfaces that you walk on. They only show up in elevation and section views. Many elements are placed relative to a *Level*; when the *Level* is moved those elements move with it (e.g., doors, windows, casework, ceilings). ***WARNING:*** *If a Level is deleted, those same "dependent" elements will also be deleted from the project!*

- **Reference Planes**
  - ❖ These are similar to grids in that they show up in plan and elevation or sections. They do not have reference bubbles at the end like grids. Revit breaks many tasks down into simple 2D tasks which result in 3D geometry. *Reference Planes* are used to define 2D surfaces on which to work within the 3D model. They can be placed in any view, either horizontally or vertically.

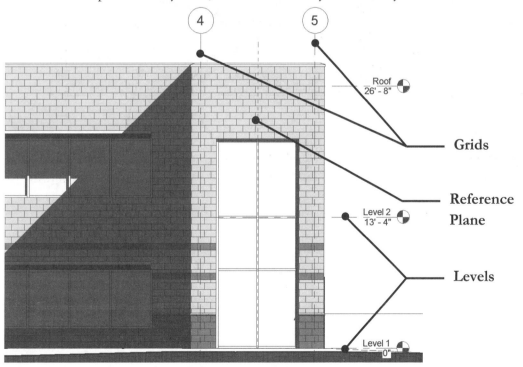

View-Specific Elements

As the name implies, the items about to be discussed only show up in the specific view in which they are created. For example, notes and dimensions added in the architectural floor plans will not show up in the structural floor plans. These elements are all 2D and are mainly communication tools used to accurately document the building for construction or presentations.

- **Annotation elements** (text, tags, symbols, dimensions)
  - Size automatically set and changed based on selected drawing scale

- **Details** (detail lines, filled regions, 2D detail components)

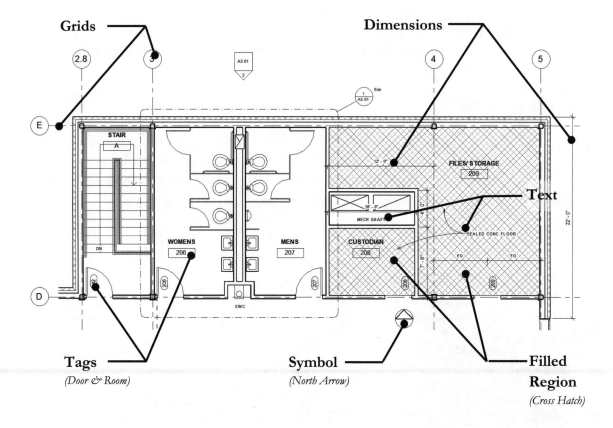

Grids

Dimensions

Text

Tags
(Door & Room)

Symbol
(North Arrow)

Filled
Region
(Cross Hatch)

The Revit platform has three fundamental ways in which to work with the elements (for display and manipulation):

- Views
- Schedules
- Sheets

The following is a cursory overview of the main ideas you need to know. This is not an exhaustive study on views, schedules and sheets.

Views

Views, accessible from the *Project Browser* (see Page 5-4), is where most of the work is done while using Revit. Think of views as slices through the building, both horizontal (plans) and vertical (elevations and sections).

- **Plans**

  A *Plan View* is a horizontal slice through the building. You can specify the location of the **cut plane** which determines if certain windows show up or how much of the stair is seen. A few examples are architectural floor plan, reflected ceiling plan, site plan, structural framing plan, HVAC floor plan, electrical floor plan, lighting [ceiling] plan, etc. The images below show this concept; the image on the left is the 3D BIM. The middle image shows the portion of building above the cut plane removed. Finally, the last image on the right shows the plan view you work in and place on a sheet.

- **Elevations**

  Elevations are vertical slices, but where the slice lies outside the floor plan as in the middle image below. Each elevation created is listed in the *Project Browser*. The image on the right is an example of a South exterior elevation view, which is a "live" view of the 3D model. If you select a window here and delete it, the floor plans will update instantly.

- **Sections**

  Similar to elevations, sections are also vertical slices. However, these slices cut through the building. A section view can be cropped down to become a wall section or even look just like an elevation. The images below show the slice, the portion of building in the foreground removed, and then the actual view created by the slice. A setting exists, for each section view, to control how far into that view you can see. The example on the right is "seeing" deep enough to show the doors on the interior walls.

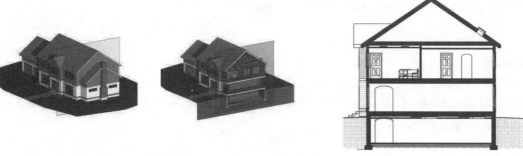

- **3D and Camera**

  In addition to the traditional "flattened" 2D views that you will typically work in, you are able to see your designs more naturally via 3D and Camera views.

  A **3D view** is simply an axonometric view; i.e., three-dimensional but without perspective.

  A **Camera view** is a true perspective view; cameras can be created both in and outside of the building. Like the 2D views, these 3D/Camera views can be placed on a sheet to be printed. Revit provides a number of tools to help explore the 3D view, such as Section Box, Steering Wheel, Temporary Hide and Isolate, and Render.

  You can use the View Cube in 3D views to toggle the view between axonometric and perspective projections

  The image on the left is a 3D view set to "shaded mode" and has shadows turned on. The image on the right is a camera view set up inside the building; the view is set to "hidden line" rather than shaded, and the camera is at eye level.

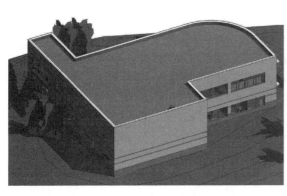

Schedules

Schedules are lists of information generated based on content that has been placed, or modeled, within the project. A schedule can be created, such as the door schedule example shown below, that lists any of the data associated with each door that exists in the project. Revit allows you to work directly in the schedule views. Any change within a schedule view is a change directly to the element being scheduled. Again, if a door were to be deleted from this schedule, that door would be instantly deleted from the project.

Think of schedules as a "non-graphical" view of model elements. You are viewing the model; you are just not seeing the graphic part of the model. You are only viewing the information about an element, not the graphic representation.

DOOR AND FRAME SCHEDULE												
DOOR NUMBER	DOOR				FRAME		DETAIL			GLAZING	FIRE RATING	HDWR GROUP	
	WIDTH	HEIGHT	MATL	TYPE	MATL	TYPE	HEAD	JAMB	SILL				
1000A	3' - 8"	7' - 2"	WD		HM		11/A8.01	11/A8.01					
1046	3' - 0"	7' - 2"	WD	D10	HM	F10	11/A8.01	11/A8.01 SIM				34	
1047A	6' - 0"	7' - 10"	ALUM	D15	ALUM	SF4	6/A8.01	6/A8.01	1/A8.01 SIM	1" INSUL		2	CARD READER N. LEAF
1047B	8' - 0"	7' - 2"	WD	D10	HM	F13	12/A8.01	11/A8.01 SIM			60 MIN	85	MAG HOLD OPENS
1050	3' - 0"	7' - 2"	WD	D10	HM	F21	8/A8.01	11/A8.01		1/4" TEMP		33	
1051	3' - 0"	7' - 2"	WD	D10	HM	F21	8/A8.01	11/A8.01		1/4" TEMP		33	
1052	3' - 0"	7' - 2"	WD	D10	HM	F21	8/A8.01	11/A8.01		1/4" TEMP		33	
1053	3' - 0"	7' - 2"	WD	D10	HM	F21	8/A8.01	11/A8.01		1/4" TEMP		33	
1054A	3' - 0"	7' - 2"	WD	D10	HM	F10	8/A8.01	11/A8.01		1/4" TEMP	-	34	
1054B	3' - 0"	7' - 2"	WD	D10	HM	F21	8/A8.01	11/A8.01		1/4" TEMP	-	33	
1055	3' - 0"	7' - 2"	WD	D10	HM	F21	8/A8.01	11/A8.01		1/4" TEMP	-	33	
1056A	3' - 0"	7' - 2"	WD	D10	HM	F10	9/A8.01	9/A8.01			20 MIN	33	
1056B	3' - 0"	7' - 2"	WD	D10	HM	F10	11/A8.01	11/A8.01			20 MIN	34	
1056C	3' - 0"	7' - 2"	WD	D10	HM	F10	20/A8.01	20/A8.01			20 MIN	33	
1057A	3' - 0"	7' - 2"	WD	D10	HM	F10	8/A8.01	11/A8.01			20 MIN	34	
1057B	3' - 0"	7' - 2"	WD	D10	HM	F30	9/A8.01	9/A8.01		1/4" TEMP	20 MIN	33	
1058A	3' - 0"	7' - 2"	WD	D10	HM	F10	9/A8.01	9/A8.01			-	33	

Sheets

You can think of sheets as the pieces of paper on which your views and schedules will be printed. Views and schedules are placed on sheets and then arranged. Once a view has been placed on a sheet, its reference bubble is automatically filled out and that view cannot be placed on any other sheet. The setting for each view, called "view scale," controls the size of the drawing on each sheet; view scale also controls the size of the text, tags and dimensions.

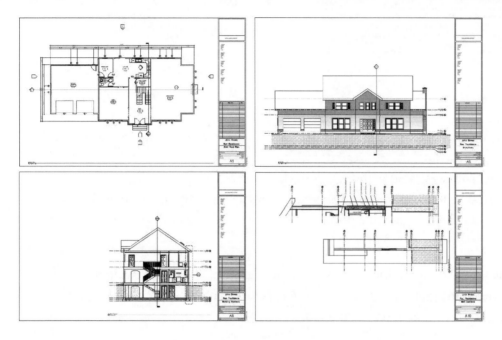

Conclusion

With this information on the fundamental concepts Revit is built upon, it is easy to learn more about how to use the software efficiently. The knowledge gained, or refreshed, from this chapter will also aid in navigating the Certification Exam.

3.0 User Interface Review

Prior to taking the Autodesk Certified User exam it is important to have a good understanding of the layout and the terms used for the user interface (UI). This will help avoid confusion if an exam question uses a UI term such as "Options Bar". It will also allow you to be more efficient during the exam in the in-application portion.

This section will walk through the different aspects of the User Interface.

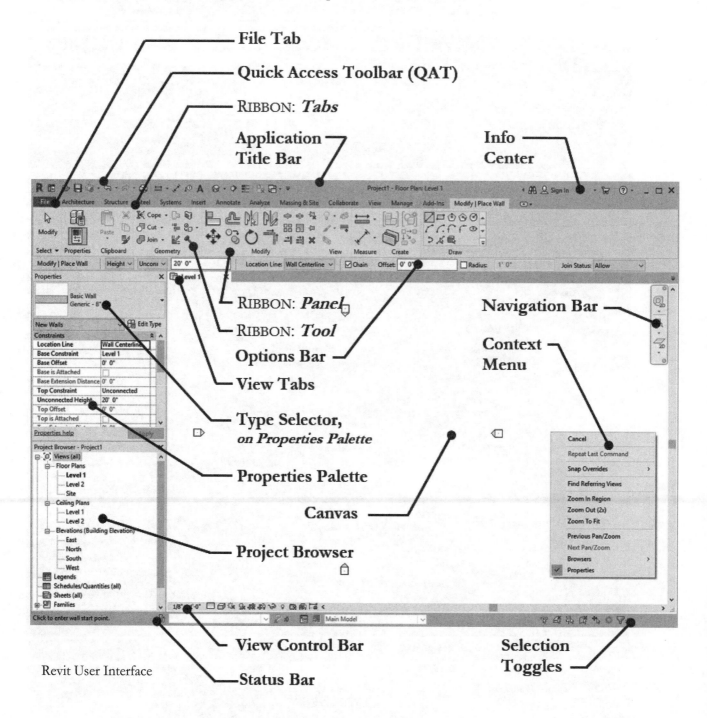

Revit User Interface

Application Title Bar:

In addition to the *Quick Access Toolbar* and *Info Center*, which are all covered in the next few sections, you are also presented with the product name, version and the current file-view in the center. As previously noted already, when working on a project, the version is important as you do not want to upgrade unless you have coordinated with other staff and/or consultants; everyone must be using the same version of Revit.

File Tab:

Access to *File* tools such as *Save*, *Plot*, *Export* and *Print* (both hardcopy and electronic printing). You also have access to tools which control the Revit application as a whole, not just the current project, such as *Options* (see the end of this section for more on *Options*).

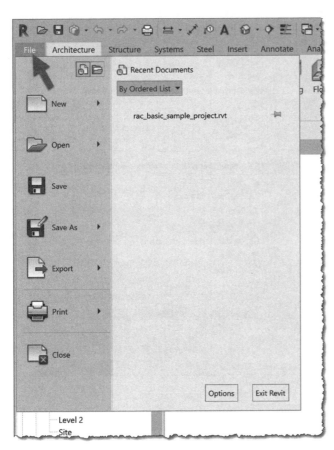

Recent and Open Documents:

These two icons (from the *File Tab*) toggle the entire area on the right to show either the recent documents you have been in (icon on the left) or a list of the documents you currently have open.

In the *Recent Documents* list you click a listed document to open it. This saves time as you do not have to click *Open* → *Project* and browse for the document (*Document* and *Project* mean the same thing here). Finally, clicking the "Pin" keeps that project from getting bumped off the list as additional projects are opened.

In the *Open Documents* list, the "active" project you are working in is listed first; clicking another project switches you to that open project.

The list on the left, in the *File Tab* shown above, represents three different types of buttons: *button*, *drop-down button* and *split button*. Save and Close are simply **buttons**. Save-As and Export are **drop-down buttons**, which means to reveal a group of related tools. If you click or hover your cursor over one of these buttons, you will get a list of tools on the right. Finally, **split buttons** have two actions depending

Button

Drop-down Button

Split Button

on what part of the button you click on; hovering over the button reveals the two parts (see bottom image to the right). The main area is the most used tool; the arrow reveals additional related options.

Quick Access Toolbar:

Referred to as *QAT* in this book, this single toolbar provides access to often used tools (*Open, Save, Undo, Redo,* Print, *Measure, Tag,* etc.). It is always visible regardless of what part of the *Ribbon* is active.

The *QAT* can be positioned above or below the *Ribbon* and any command from the *Ribbon* can be placed on it; simply right-click on any tool on the *Ribbon* and select *Add to Quick Access Toolbar*. Moving the *QAT* below the *Ribbon* gives you a lot more room for your favorite commands to be added from the *Ribbon*. Clicking the larger down-arrow to the far right reveals a list of common tools which can be toggled on and off, as well as the 'Customize' dialog shown to the right.

Some of the icons on the *QAT* have a down-arrow on the right. Clicking this arrow reveals a list of related tools. In the case of *Undo* and *Redo,* you have the ability to undo (or redo) several actions at once.

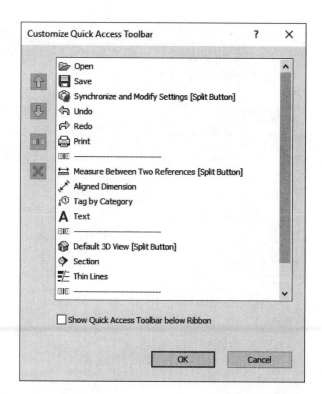

Ribbon – Architecture Tab:

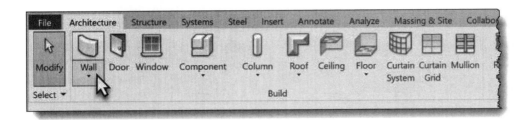

The *Architecture* tab on the *Ribbon* contains most of the tools the architect needs to model a building, essentially the things you can put your hands on when the building is done. The specific discipline versions of Revit omit some of the other discipline tabs.

Each tab starts with the *Modify* tool, i.e., the first button on the left. This tool puts you into "selection mode" so you can select elements to modify. Clicking this tool cancels the current tool and unselects elements. With the *Modify* tool selected you may select elements to view their properties or edit them. Note that the *Modify* tool, which is a button, is different than the *Modify* tab on the *Ribbon*.

The *Ribbon* has three types of buttons: *button*, *drop-down button* and *split*, as covered on the previous page. In the image above you can see the *Wall* tool is a **split button**. Most of the time you would simply click the top part of the button to draw a wall. Clicking the down-arrow part of the button, for the *Wall* tool example, gives you the option to draw a *Wall*, *Structural Wall*, *Wall by Face*, *Wall Sweep*, and a *Reveal*.

TIP: The Model Text tool is only for placing 3D text in your model, not for adding notes!

Ribbon – Annotate Tab:

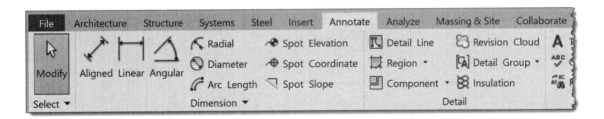

To view this tab, simply click the label "Annotate" near the top of the *Ribbon*. This tab presents a series of tools which allow you to add notes, dimensions and 2D "embellishments" to your model in a specific view, such as a floor plan, elevation, or section. All of these tools are **view specific**, meaning a note added in the first-floor plan will not show up anywhere else, not even another first-floor plan: for instance, a first-floor electrical plan.

Notice, in the image above, that the *Dimension* panel label has a down-arrow next to it. Clicking the down-arrow will reveal an **extended panel** with additional related tools.

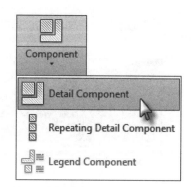

Finally, notice the *Component* tool in the image above; it is a **split button** rather than a *drop-down button*. Clicking the top part of this button will initiate the *Detail Component* tool. Clicking the bottom part of the button opens the fly-out menu revealing related tools.

Ribbon – Modify Tab:

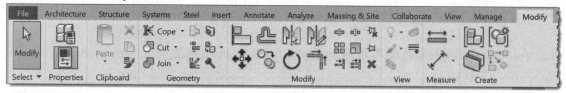

Several tools which manipulate and derive information from the current model are available on the *Modify* tab. Additional *Modify* tools are automatically appended to this tab when elements are selected in the model (see *Modify Contextual Tab* on the next page).

> *TIP: Do not confuse the Modify tab with the Modify tool when following instructions in this book.*

Ribbon – View Tab:

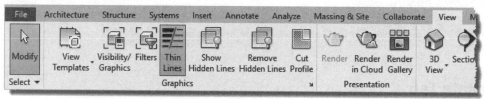

The tools on the *View* tab allow you to create new views of your 3D model; this includes views that look 2D (e.g., floor plans, elevations and sections) as well as 3D views (e.g., isometric and perspective views).

The *View* tab also gives you tools to control how views look, everything from what types of elements are seen (e.g., Plumbing Fixture, Furniture or Section Marks) to line weights.

> *NOTE: Line weights are controlled at a project wide level but may be overridden on a view by view basis.*

Finally, notice the little arrow in the lower-right corner of the *Graphics* panel. When you see an arrow like this you can click on it to open a dialog box with settings that relate to the panel's tool set (*Graphics* in this example). Hovering over the arrow reveals a tooltip which will tell you what dialog box will be opened.

Ribbon – Modify Contextual Tab:

The *Modify* tab is appended when certain tools are active or elements are selected in the model; this is referred to as a *contextual tab*. The first image below shows the *Place Wall* tab which presents various options while adding walls. The next example shows the *Modify Walls* contextual tab which is accessible when one or more walls are selected.

Place Wall contextual tab – visible when the Wall tool is active.

Modify Walls contextual tab – visible when a wall is selected.

Ribbon – States:

The *Ribbon* can be displayed in one of four states:

- Full Ribbon (default)
- Minimize to Tabs
- Minimize to Panel Tiles
- Minimize to Panel Buttons

The intent of this feature is to increase the size of the available drawing window. It is recommended, however, that you leave the *Ribbon* fully expanded while learning to use the program. The images in this book show the fully expanded state. The images below show the other three options. When using one of the minimized options you simply hover (or click) your cursor over the Tab or Panel to temporarily reveal the tools.

FYI: Double-clicking on a Ribbon tab will also toggle the states.

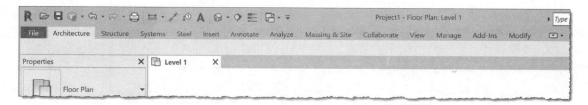

Minimize to Tabs

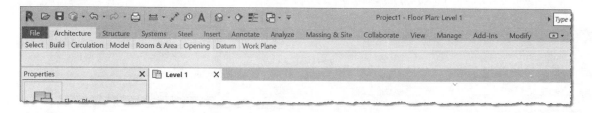

Minimize to Panel Tiles

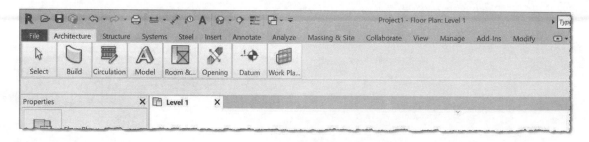

Minimize to Panel Buttons

Options Bar:

This area dynamically changes to show options that complement the current operation. The *Options Bar* is located directly below the *Ribbon*. When you are learning to use Revit you should keep your eye on this area and watch for features and options appearing at specific times. The image below shows the *Options Bar* example with the *Wall* tool active.

Properties Palette – Element Type Selector:

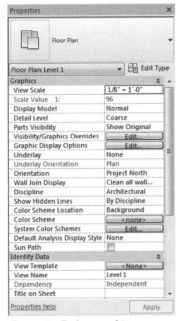

Properties Palette; nothing

The *Properties Palette* provides instant access to settings related to the element selected or about to be created. When nothing is selected, it shows information about the current view. When multiple elements are selected, the common parameters are displayed.

The *Element Type Selector* is an important part of the *Properties Palette*. Whenever you are adding elements or have them selected, you can select from this list to determine how a wall to be drawn will look, or how a wall previously drawn should look (see image to left). If a wall type needs to change, you never delete it and redraw it; you simply select it and pick a new type from the *Type Selector*.

The **Selection Filter** drop-down list below the *Type Selector* lets you know the type and quantity of the elements currently selected. When multiple elements are selected you can narrow down the properties for just one element type, such as *wall*. Notice the image to the left shows four walls are in the current selection set. Selecting **Walls (4)** will cause the *Palette* to only show *Wall* properties even though many other elements are selected (and remain selected).

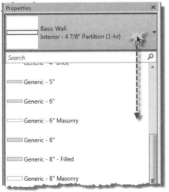

Type Selector; Wall tool active or a Wall is selected

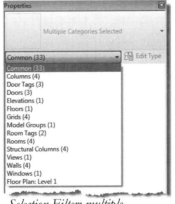

Selection Filter; multiple elements selected

The width of the *Properties Palette* and the center column position can be adjusted by dragging the cursor over that area. You may need to do this at times to see all the information. However, making the *Palette* too wide will reduce the useable drawing area.

The *Properties Palette* should be left open; if you accidentally close it you can reopen it by **View → Window → User Interface → Properties** or by typing **PP** on the keyboard.

Project Browser:

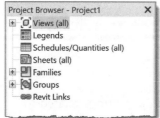

The *Project Browser* is the "Grand Central Station" of the Revit project database. All the views, schedules, sheets and content are accessible through this hierarchical list. The first image to the left shows the seven major categories; any item with a "plus" next to it contains sub-categories or items.

Double-clicking on a View, Legend, Schedule or Sheet will open it for editing; the current item open for editing is bold (**Level 1** in the example to the left). Right-clicking will display a pop-up menu with a few options such as *Delete* and *Copy*.

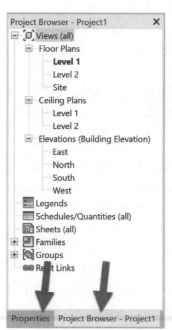

Right-click on *Views (all)*, at the top of the *Project Browser*, and you will find a **Search** option in the pop-up menu. This can be used to search for a *View*, *Family*, etc., a very useful tool when working on a large project with 100s of items to sift through.

Like the *Properties Palette*, the width of the *Project Browser* can be adjusted. When the two are stacked above each other, they both move together. You can also stack the two directly on top of each other; in this case you will see a tab for each at the bottom as shown in the second image to the left.

The *Project Browser* should be left open; if you accidentally close it by clicking the "X" in the upper right, you can reopen it by **View → Window → User Interface → Project Browser**.

The *Project Browser* and *Properties Palette* can be repositioned on a second monitor, if you have one, when you want more room to work in the drawing window.

> If the **Project Browser** or **Properties Palette** are accidentally closed, open it via **View → User Interface**—checked options here are open/visible.

View Tabs

Revit displays a tab for each open view. Clicking a tab is a quick way to switch between open views. Click the "X" in each view tab to close that view. Drag a tab to change its position. As you drag a view to a different position, a preview of the view position will be shown before you dock it in place. Tabs can also be pulled outside of the main Revit application window – even to a second monitor. In the second image below, the schedule could be filling a second computer screen while reviewing the same information in the floor plan.

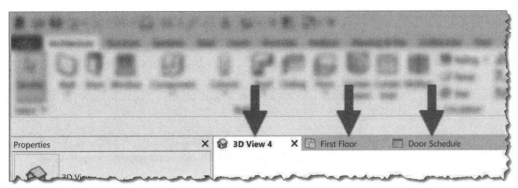

View Tabs; one for each open view

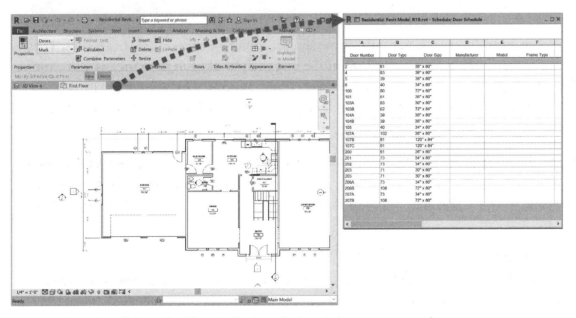

View Tabs can be pulled outside of the main Revit application window, even to a second screen

Status Bar:

This area will display information, on the far left, about the current command or list information about a selected element. The right-hand side of the *Status Bar* shows the number of elements selected. The small funnel icon to the left of the selection number can be clicked to open the *Filter* dialog box, which allows you to reduce your current selection to a specific category; for example, you could select the entire floor plan, and then filter it down to just the doors. This is different than the *Selection Filter* in the *Properties Palette* which keeps everything selected.

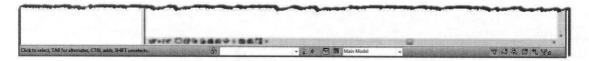

On the *Status Bar*, the five icons on the right, shown in the image below, control how elements are selected. These are, from left to right:

- Select Links
- Select Underlay Elements
- Select Pinned Elements
- Select Elements by Face
- Drag Elements on Selection

Hover your cursor over an icon for the name and for a brief description of what it does. These are toggles that are on or off; **the red 'X' in the upper right of each icon means you cannot select that type of element within the model.** These controls help prevent accidentally moving or deleting things. Keep these toggles in mind if you are having trouble selecting something; you may have accidentally toggled one of these on.

Finally, the two drop-down lists towards the center of the *Status Bar* control **Design Options** and **Worksets** (see image on previous page). The latter is not covered in this book, but *Design Options* are. *Worksets* relate to the ability for more than one designer to be in the model at a time.

Escape Key:

Pressing the Esc key in Revit will unselect an element and/or cancel the current command. Sometime this key needs to be tapped two – three times when in sub-commands. Note, when in a sketch mode, e.g. creating a floor, pressing the Esc key does not cancel the command. Rather, the red "X" or green "checkmark" must be used to cancel or finish the command.

View Control Bar:

This is a feature which gives you convenient access to tools which control each view's display settings (i.e., scale, shadows, detail level, graphics style, etc.). The options vary slightly between view types: 2D View, 3D view, Sheet and Schedule. The important thing to know is that these settings only affect the current view, the one listed on the *Application Title Bar*. Most of these settings are available in the *Properties Palette*, but this toolbar cannot be turned off like the *Properties Palette* can. Some of the tools act as a "mode" for the view. When a "mode" is turned on, the button will be highlighted on the view control bar, and the border of the view will be highlighted in a similar color.

Context Menu:

The *context menu* appears near the cursor whenever you right-click on the mouse (see image at right). The options on that menu will vary depending on what tool is active or what element is selected.

Canvas:

This is where you manipulate the Building Information Model (BIM). Here you will see the current view (plan, elevation or section), schedule or sheet. Any changes made are instantly propagated to the entire database.

Context menu example with a wall selected

Elevation Marker:

This item is not really part of the Revit UI but is visible in the drawing window by default via the various templates you can start with, so it is worth mentioning at this point. The four elevation markers point at each side of your project and ultimately indicate the drawing sheet on which you would find an elevation drawing of each side of the building. All you need to know right now is that you should draw your floor plan generally in the middle of the four elevation markers that you will see in each plan view; DO NOT delete them as this will remove the related view from the *Project Browser*.

Notes:

4.0 Creating & Modifying Components

Create and Modify Grids

Grids are used to coordinate and define the location of major structural elements within a building, such as foundations, columns and beams. They also become a communication tool during the design and construction phase of a project.

> For elevation and section views, Grids only appear if they intersect the cut plane of the view and are perpendicular to the view. Thus, Grids curved in plan never show up in elevation/section views.

There are two things to keep in mind while creating grids.

The **first** is to pick the Grid endpoints in the correct order. If you pick backwards, meaning the grid bubble is on the opposite side from what you wanted, you should click Undo and pick the points in the opposite order. It is possible to toggle the grid head so that the bubble is on the correct side without using Undo, but that change is an override and only applies to the current view. Thus, all other views, including the consultant models, will be backwards.

The **second** is to align the endpoints of the Grid being created with the endpoints of previously drawn grids. When this is done they will be constrained and move with each other. Simply watch for the green dashed line before clicking each endpoint.

A common issue is that a Grid line is adjusted in an elevation view so the line does not extend through the drawing. However, simply dragging the endpoint upward changes the physical extents of the line such that it does not engage the floor plan cut plane, which means this grid will not appear in a plan view. The proper method to adjust the elevation view is to select the Grid, click the 3D icon, and toggle it to say 2D, on the end to be adjusted. Now dragging the endpoint only affects the current view.

A new grids number/letter is based on the previous one. Two grids cannot have the same number or letter. The grid head can be displayed on both ends of the grid line if desired.

quick steps

Grid

1. Ribbon
   - Line Segment
   - or Pick Line
   - Multi-Segment
2. Options Bar
   - Offset
3. Type Selector
   - Select Grid Type
4. Properties
   - Scope Box
5. In-Canvas
   - Pick two points to define Grid
     - First point is end without symbol – this will be the default for all views
   - When ends align during placement, the new Grid will adjust in length with adjacent Grids.

Create and Modify Levels

Levels are horizontal reference planes which typically define a surface a person can walk on. Graphically, Levels have a start and end point, but technically they are infinite.

Levels are only visible, and can only be created, in elevation or section views. In Revit 2019 and beyond, Levels also appear in 3D views. Also, Levels only appear in views they intersect.

> Prior to Revit 2019, **Deleting a Level** will delete all elements associated with it – and Revit does not provide any warnings, so delete Levels with care!

To change the **Elevation** of a **Level**, select it in an elevation or section view and edit the elevation value via the Properties Palette as shown in the image below.

Creating a Level creates three views: Floor Plan, Ceiling Plan and Structural Plan, based on default Plan View Types on Options Bar. Copying a level does not create views automatically.

Since most elements in Revit are associated with a level, directly or indirectly, changing the elevation causes many elements to move.

Level

1. Must be in elevation or section view
2. Ribbon
   - Sketch or Pick Line
3. Options Bar
   - Make Plan View
   - Plan View Types…
   - Offset
4. Type Selector
   - Select Level Type
5. Properties
   - Story Above
   - Computation Height
   - Scope Box
   - Structural
   - Building Story
6. Type Properties
   - Elevation Base
7. In-Canvas
   - Pick two points to define level
     - First point is ended without symbol
   - When ends align during placement, the new Level will adjust in length with adjacent Levels.

Create and modify walls

The Wall tool is used to model interior and exterior walls, curtain walls (i.e., glass walls), foundation walls and retaining walls. In the context of ceiling design, walls are also used to develop bulkheads and the vertical portion of a soffit.

Initiating the Wall command is accomplished by simply clicking the top portion of the Wall tool on the Architecture tab.

There are multiple "**Offset**" parameters. Offset on the *Options Bar* is for horizontal displacements, and Base Offset in *Properties* is vertical relative to the floor level.

By default, walls are **Room Bounding** (controlled by an instance parameter). Adding a wall to a room that already has a Room element placed causes the Room element to readjust, which affects the Room's **Area** and **Volume** values. Pay attention to the **Location Line** selection on the *Options Bar*; if the thickness of the wall is on the wrong side, that will negatively affect things like length, area, volume, etc. which may be on the exam. Windows and doors **subtract area** for the wall, as shown on the next page.

> Pressing the **Spacebar** while adding a wall will flip the wall along the sketch line, making the exterior side of the wall switch to the other side.

Know how to use the modify commands **Trim/Extend to Corner** & **Trim/Extend Single Element** on walls. With a wall selected, click Edit Type in Properties, and then **Duplicate** to make another wall type. Modify the Type Parameter **Structure** via the Edit button to add or remove layers of construction within the wall. Sloped walls are achieved by setting the wall's **Cross-Section** parameter to Slanted, and then adjusting the **Angle From Vertical** value.

Wall: Architectural

1. Ribbon
   - Draw panel
   - *Default:* straight segment
2. Options Bar
   - Level (3D Views)
   - Height
   - Location Line
   - Chain
   - Offset
   - Radius
   - Join Status
3. Type Selector
   - Wall Type
     - Basic Wall
     - Curtain Wall
     - Stacked Wall
4. Properties
   - Base Constraint
   - Base Offset
5. In-Canvas
   - Spacebar
   - Snaps
   - Temporary Dimensions

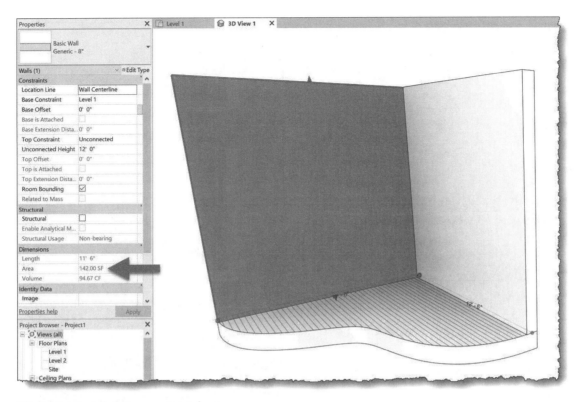

Wall selected, and surface area pointed out

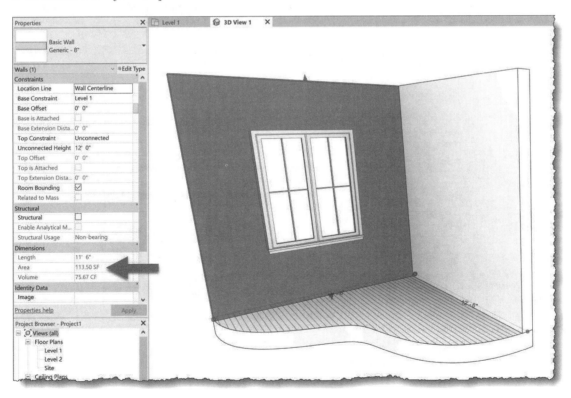

Wall with window selected, and reduced surface area pointed out

Load and modify doors

The Door tool is used to model interior and exterior doors.

Door

Initiating the Door command is accomplished by simply clicking the command on the Architecture tab via the Ribbon.

Doors are wall hosted, can be placed in plan, elevation or 3d views and can be deleted from within a schedule.

Note the following for the selected door in the image below:

1. **Temporary dimensions** used to accurately position door
2. **Flip controls** to change door swing and hand
3. **Type Selector**; change door type for selected door
4. **Level** door is associated with and vertical **offset** if needed

quick steps

Door

1. Ribbon
   - Load Family
   - Model In-Place
   - Tag on Placement
2. Type Selector
   - Select Door Type
3. Options Bar
   - Tag options
4. Properties
   - Sill/Head Height
   - See Warning
5. In-Canvas
   - Hover over wall
   - Swing Direction
       i. Hover cursor near swing side of wall
   - Hinge Side
       i. Spacebar to flip

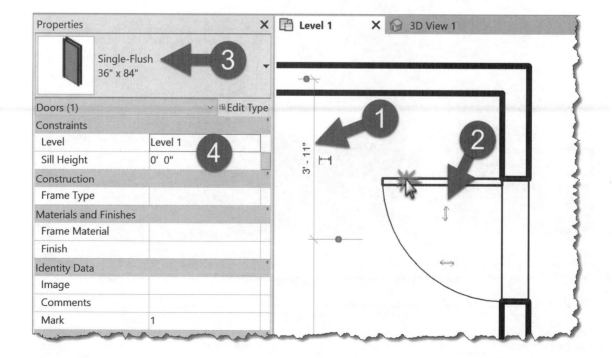

Load and modify windows

The Window tool is used to model interior and exterior windows.

Initiating the Window command is accomplished by simply clicking the command on the Architecture tab via the Ribbon.

By default, the Sill Height is an instance parameter, while Width and Height are type parameters.

Note the following for the selected door in the image below:

1. **Temporary dimensions** used to accurately position window
2. **Flip controls** to change window orientation within wall
3. **Type Selector**; change window type for selected window
4. **Level** window is associated with and **Sill Height** (vertical **offset)**

Window

1. Ribbon
   - Load Family
   - Model In-Place
   - Tag on Placement
2. Options Bar
   - Rotate after placement
   - Level (in 3D views)
3. Type Selector
   - Select Door Type
4. Properties
   - Sill/Head Height
   - See Warning
5. In-Canvas
   - Hover over wall
   - Swing Direction
     i. Hover cursor near swing side of wall
   - Hinge Side
     i. Spacebar to flip

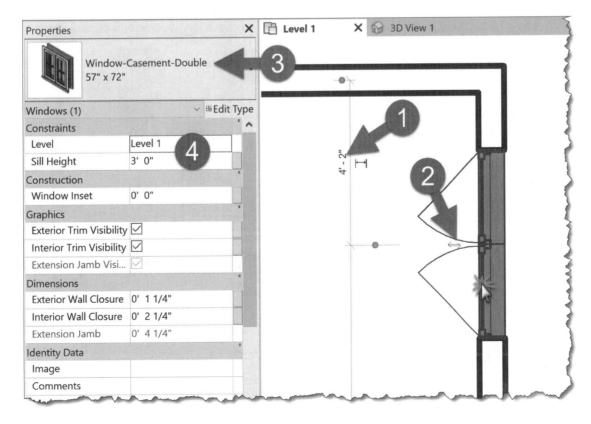

Tag components by category

This tool, often just referred to as the Tag command, is as significant as the Wall tool in the world of Revit—which is why it is also located on the *Quick Access Toolbar*. This tool is used to graphically display information contained within model elements. For example, a tag can be used to display a door's number in a floor plan, or a window type in an elevation view. If the door or window is changed, via the Type Selector, the tag automatically updates. The tag itself holds no information; it just reports information from the element it is tagging.

Each category in Revit, i.e., Door, Window, Wall, Furniture, Column, etc., must have its own tag family. Revit allows the user to specify which tag is used, by default, for each category. The tag may be changed at any time via the Type Selector. A single element may be tagged more than once. You might tag the same door in the main floor plan and an enlarged floor plan. Another example might be tagging the same element, in the same view, with more than one tag; a light fixture may have three different tags: circuit number, fixture type and switch system.

The image to the right has several tags added to a floor plan view. All the listed information is coming from the properties of the elements which have been tagged. For example, the "M1" tag within the diamond shape is listing the wall type (i.e., *Type Mark*). Because the wall tag(s) is/are listing a *Type Parameter*, all wall instances of that type will report the same value, i.e., "M1." The door number "1" is reporting the element's **Mark** value, which is an *Instance*

quick steps

Tag by Category

1. Options Bar
    *   Tag orientation
        i.   Horizontal
        ii. Vertical
    *   Tags… (button)
    *   Leader
    *   Leader type
        i.   Attached End
        ii. Free End
    *   Leader Length
2. In-Canvas
    *   Hover cursor over model elements, click to place tag
    *   3D Views must be locked before placing tags
    *   Notice the specific element you are about to tag is listed on the status bar in the lower left of the screen

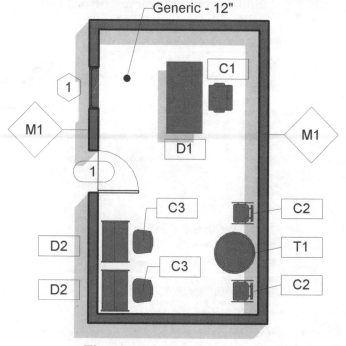

Floor plan with tags added

Parameter. Therefore, each door instance may have a different number. Notice some tags have the *Leader* option turned on. This is especially helpful if the tag is outside the room. Any 3D element visible in a view may be tagged, even the floor. In this case the *Floor Tag* is actually reporting the *Type Name* listed in the *Type Selector*. The leader can be modified to have an arrow, a dot or nothing.

All of these tags were placed using the same tool: **Tag by Category**.

The section / interior elevation below shows many of the same elements tagged as were tagged in the floor plan view above. If the wall's *Type Mark* is changed, the wall *Tag* will be instantly updated in all views: plans, elevations, sections, and schedules.

Some tags can actually report multiple parameter values found within an element. For example, the **Ceiling Tag** used in this example lists the *Type Name*, the ceiling height, and has fixed text which reads "A.F.F." Many tags can be selected and directly edited, which actually changes the values within the element. Changing the 8'-0" ceiling height will cause the ceiling position to change vertically.

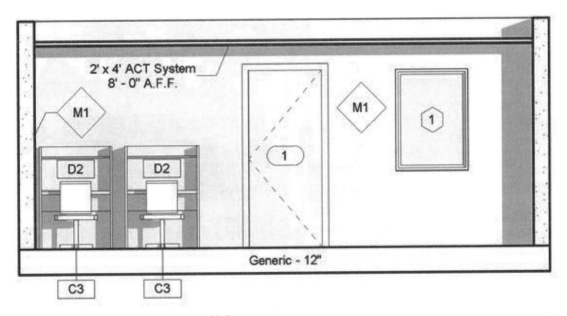

Section / Interior Elevation with tags added

Load and modify components

If an element does not have its own tool (e.g. Door, Window, Ceiling) then it is placed using the catch-all **Component** tool.

Note the following in the image below:

1. **Temporary dimensions** used to accurately position element
2. **Type Selector**; change type for selected element
3. **Level** element is associated with

The desk and chair are Components in the image below.

With a Component selected, its properties can be seen in the Properties Palette (instance properties) and by clicking Edit Type (type properties).

quick steps

Place a Component

1. Ribbon
   - Load Family
   - Model In-Place
2. Options Bar
   - Rotate after Placement
3. Type Selector
   - Select Family to place
4. Properties
   - Options vary
   - See Warning
5. In-Canvas
   - Spacebar to rotate
      i. Cursor location

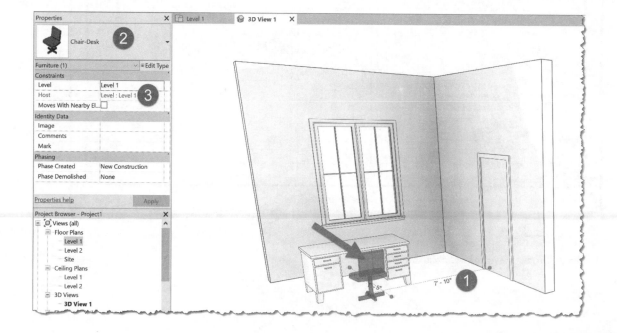

5.0 Modeling & Modifying Elements

Create a roof & modify roofs

The **Roof by Footprint** tool is used to draw an outline of the roof in plan view. The lines, which represent the perimeter of the roof, may define a sloped surface (see Defines Slope in Options Bar section below).

Picking this tool enters a **Sketch Mode** where all elements are grayed out and not selectable. The goal is to create a closed outline with no gaps or overlapping lines. Any secondary outlines within a larger outline define a hole.

> The only way to get out of the Roof tool is to click the **green check mark** (finish) or the **red X** (cancel) on the Ribbon.

Know the related options while creating a roof: Options Bar, Type Selector & Properties, as shown here:

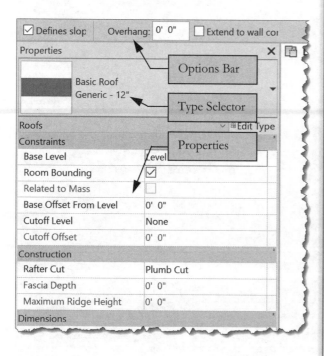

quick steps

Roof by Footprint

1. First Switch to Appropriate Plan View
   - i.e., View associated with the level the roof should be aligned with
2. Ribbon [in sketch mode]
   - Draw *Panel*
     - Boundary Line
     - Slope Arrow
     - Segment or Pick Options
   - Work Plane *Panel*
     - Set/Show Work Plane
     - Create Reference Plane
       i. Tip: Name it
     - Viewer
   - Tools *Panel*
     - Align Eaves
3. Options Bar
   - Defines Slope
     - Line defines the edge of a sloped roof plane
   - Overhang
   - Extend to core
4. Type Selector
   - Select Roof Type
5. Properties (three groups)
   - *Roofs*
     - Level settings
   - *Sketch Lines*
     - Set pitch
   - *View*
6. Green Checkmark to Finish Sketch
   - Must have enclosed footprint with no overlapping lines or gaps at intersections
7. In-Canvas
   - Select line to adjust pitch or Defines Slope Setting
   - Select Slope Arrow to adjust settings

When sketching the perimeter of a roof, if **Defines Slope** is checked (see Options Bar on previous page) the roof edge defined by that line will slope. While in Sketch Mode, a small right-triangle appears next to each line that defines a slope as seen in the image below. **Select the sketch line to change the slope via properties.** Here, there are three buildings with the same perimeter roof sketch. Notice how the result is different depending on which lines define a slope. The Defines Slope setting can be changed by editing the roof sketch, selecting a sketch line and then un-checking defines slope via Properties.

Tip: The area, volume and perimeter of a selected roof can be seen in the Properties Palette.

Some roof forms cannot be sketched in a plan view, such as a barrel vault roof, so Revit provides the **Roof by Extension** tool. Using this tool, the top edge of the roof form is sketched in a non-plan view, e.g. elevation, section, 3D.

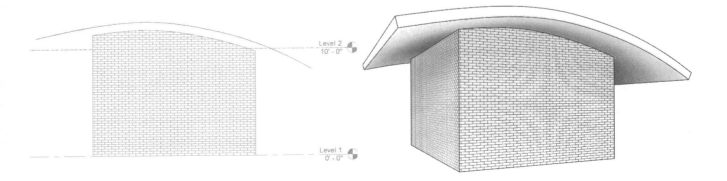

Create and modify stairs

Stair (originally Stair by Component) allows the designer to pick two points in the model and instantly see a run of 3D stairs added between them.

Stairs are a **System Family** and can only exist within a project. Thus, a library of custom options must exist in a project template or copy/pasted from another project file.

Starting at the lower level, select a point to start the stair. The image below shows what we see prior to picking our second point. Notice the dashed line assures a horizontal position and the text below indicates the number of risers created and the number remaining if we clicked at that position. Creating a stair run short of the Top Level will result in a landing being automatically added if we continue to click points to draw additional stair runs, otherwise we get a warning if the stair is finished before reaching the Top level.

All risers are the same height in a run of stairs.

Once a stair is finished/created, the dimension tool can be used to discover a distance from adjacent elements to various parts of the stair.

To modify the stair later, simply select it and pick **Edit Stair** on the Ribbon. The same methods used to create the stair can be used to modify it.

quick steps

Stair

1. Ribbon [in sketch mode]
   - Components *Panel*
     - Select from:
       i. Run (default)
       ii. Landing
       iii. Support
     - Stair shape (in plan)
   - Work Plane *Panel*
     - Set/Show Work Plane
     - Create Reference Plane
     - Viewer
   - Tools *Panel*
     - Railing (optional)
2. Options Bar
   - Location Line
   - Offset
   - Width
   - Automatic Landing
3. Type Selector
   - Select Stair Type
4. Properties
   - Base/Top Level settings
   - Actual Tread Depth
5. Type Properties (via Edit Type)
   - Maximum Riser Height
   - Minimum Tread Depth
   - Several "type" options
6. In-Canvas
   - Pick to start stair run
   - Number of risers created shown as cursor is moved

First, let's define some terms related to stairs and railings in Revit. These terms generally correspond to real-world architecture/construction terminology. However, the main goal here is to define the terms in the context of Revit.

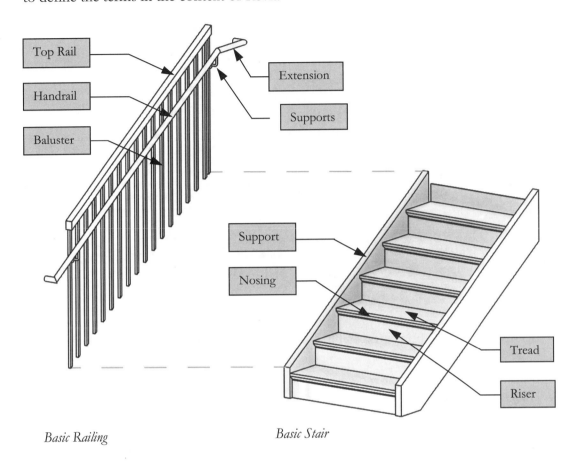

Basic Railing

Basic Stair

- Revit Stair Element
    - **Run**: a continuous section of stairs, consisting of Risers and Treads, between main floor levels and/or landings.
    - **Tread**: the flat horizontal part you step on.
    - **Riser**: the vertical portion which fills the gap between the treads. Not all stairs have risers.
    - **Nosing**: the outer edge, where the riser and tread come together.
    - **Landing**: intermediate horizontal surface (i.e. floor) between the main floor levels of a building. Building codes require a landing if a stair "run" rises more than a certain distance—which allows someone to have a safe place to rest.
    - **Supports** (aka Stringer): the structural elements supporting the treads and risers. For a commercial project this is typically a steel tube or C-channel. A stair typically has a stringer on each side of the stair. For wider stairs, one or more stringers may be required. They either span between floors (e.g. Level 1 up to Level 2) or may be anchored to an adjacent wall.

Here are the basic steps for the Stair tool:

- Start at the lowest level
- Select **Stair** tool
- Verify stair style in **Type Selector**
- Verify Base and Top Level settings in **Properties**
- Adjust settings on **Options Bar**
  - Location Line
  - Offset (horizontal)
  - Stair Width
  - Landing creation
  - See first image below
- **Pick points** to define one or more runs
- *Optional:* Add **Landing** to top or bottom if needed
  - select *Landing* on Ribbon while in Stair by Component tool
- *Optional:* Select and delete individual **Support** elements
  - For example, if there is a door at an intermediate landing
  - Use the *Support* option to replace if needed

The image below shows the **Options Bar** while the Stair by Component tool is active (see image below). Be sure to select the best option for Location Line. If in a stair shaft, one of the exterior support options is likely the more convenient option. Note that these settings can vary per run within the same stair instance.

Options bar for Stair by Component tool

Two clicks could create a continuous single run from floor to floor, or, as seen in the image below, multiple clicks create multiple runs. Landings are automatically added between runs. All along, Revit indicates how many risers remain before reaching the next level.

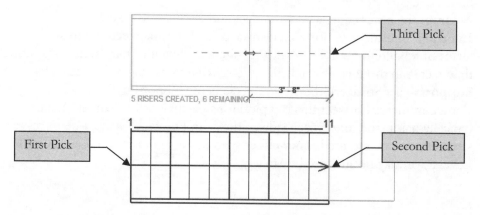

Creating stair with landing

Create and modify ramps

The steps to create a ramp are very similar to creating a stair.

Pick two points in the model and instantly see a 3D ramp added between them.

Ramps are a **System Family** and can only exist within a project. Thus, a library of custom options must exist in a project template or copy/pasted from another project.

Starting at the lower level, select a point to start the ramp. Creating a ramp run short of the Top Level will result in a landing being automatically added if we continue to click points to draw additional ramp runs, otherwise we get a warning if the ramp is finished before reaching the Top level.

Once a ramp is finished/created, the dimension tool can be used to discover a distance from adjacent elements to various parts of the stair.

To modify the ramp later, simply select it and pick **Edit Sketch** on the Ribbon. The same methods used to create the ramp can be used to modify it.

quick steps

Ramp

1. Ribbon [in sketch mode]
   - Components *Panel*
     - Select from:
       i. Run (default)
       ii. Boundary
       iii. Riser
   - Work Plane *Panel*
     - Set/Show Work Plane
     - Create Reference Plane
     - Viewer
   - Tools *Panel*
     - Railing (optional)
2. Options Bar
   - N/A
3. Type Selector
   - Select Ramp Type
4. Properties
   - Base/Top Level settings
   - Width
5. Type Properties (via Edit Type)
   - Shape (see image below)
     - Thick
     - Solid
   - Thickness
   - Function
     - Interior
     - Exterior
6. In-Canvas
   - Pick to start ramp run
   - Length of inclined ramp created/remaining shown

Level 1
0' - 0"

Level 1
0' - 0"

Create and modify railings

The Railing tool is used to place a railing in the model. Railings are added at floor edges in atriums, at ramps and stairs. Railings can also be hosted to a topographic surface (e.g. fencing).

The Stair and Ramp tools have the option of placing a railing automatically.

> The style of a selected railing can be changed in the Type Selector.

Railings have a special Handrail and Top Rail which can be selected separately. These special sub-elements have their own type properties.

The direction the railing is drawn, e.g. from left to right, determines the orientation; click the arrows while in sketch mode to swap start and end definition. Selecting the railing after it is created reveals a flip icon to change the orientation if needed.

> The height of a railing, and handrail, is defined in its Type Properties.

When a railing is selected, its total length is listed in the Properties Palette.

Railing: Sketch Path

1. Ribbon [in sketch mode]
   - Draw *Panel*
     - Sketch connected path of railing
     - Segment or Pick Options
   - Work Plane *Panel*
     - Set/Show Work Plane
     - Create Reference Plane
     - Viewer
   - Tools *Panel*
     - Pick New Host
     - Edit Joins
   - Options *Panel*
     - Preview
2. Options Bar
   - Chain
   - Offset
   - Radius
3. Type Selector
   - Select Railing Type
4. Properties
   - Base Level
   - Base Offset
   - Tread/Stringer Offset
5. In-Canvas
   - Typically work in plan view
   - Sketch connected path
   - Path cannot close back on itself

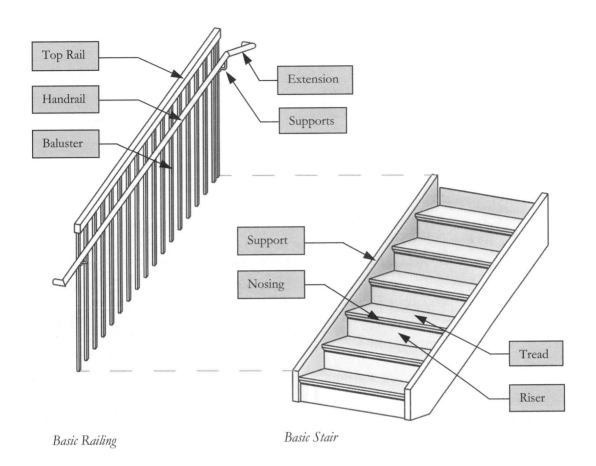

Basic Railing

Basic Stair

- <u>Revit Railing Element</u>
  - o **Balusters**: vertical elements extending from floor or stair up to "top rail." For commercial projects, building codes state that a 4" sphere cannot pass through a railing system. **FYI**: Balusters are not needed if glass panels are used.
  - o **Handrails**: continuous rounded element, attached to a wall or balusters with a "support," along a floor edge, a ramp or stair "run" which allows someone to place their hand on to prevent falling.
  - o **Guardrail**: A taller railing system, which may include a handrail, to prevent falling from a stair or floor edge. **FYI**: In the USA, when a guardrail is required it must be at least 42 inches tall.
  - o **Supports**: bracket used to attach handrail to wall or balusters.
  - o **Top Rails**: continuous rail at the top of the railing system—supported by balusters.

There are two main ways a railing can be added to a Revit project:

- In conjunction with the **Stair** or **Ramp** tools
- Separately, using the **Railing** commands
  - a. Sketch Path
  - b. Place on Host

There are two main ways a railing can be added to a Revit project:

- In conjunction with the **Stair** or **Ramp** tools
- Separately, using the **Railing** commands
  a. Sketch Path
  b. Place on Host

Railing: Place on Host

When a new project is created using one of the default Revit templates, and a stair (or Ramp) is created, a railing is automatically placed on both sides. The image below shows what the default results look like in a 3D view. These two railings are hosted by the stair and will update if the stair is modified. There are a number of things we can change to convey the design intent when needed.

If all railings have been deleted, use the **Railing → Place on Host** tool to quickly recreate the railings hosted by the stair. This tool only works if all railings have been deleted from the stair.

Railing: Sketch Path

When a railing needs to be created apart from a stair or ramp, use the **Railing → Sketch Path** tool. When this command is selected, you simply sketch a path for the railing system to follow. If the railing has a Handrail, the side it is placed on is determined by which direction the sketch is created. While in sketch mode, click the **Preview** option to see the railing as it is to be sketched (similar to how Stair by Component works).

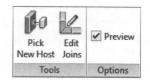

Once created, when a railing is selected a fill arrow appears as shown below. Clicking this icon flips the entire railing about the sketch line. Use this if the handrail is positioned on the wrong side of the railing. Also, the total railing length is listed in the Properties Palette.

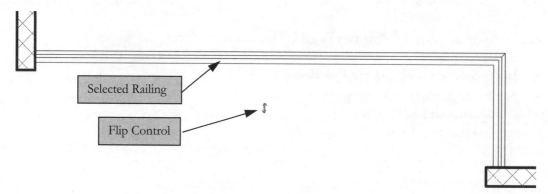

Add and modify floors

The floor tool is used to create a floor element relative to the level associated with the current plan view. The Floor may consist of one or more layers of materials. Sometimes the floor's overall thickness includes space for the structure (e.g. joists or beams). However, the Floor tool is not able to show any actual structural elements—they must be modeled separately with the Structural Framing tool.

While creating a floor, the goal is to sketch the perimeter of the floor. Any areas sketched within the largest perimeter will define a hole in the floor. The edge of the slab is always flat. Use the separate Slab Edge tool to add a slope or a thickened edge.

When a floor is created, it is associated with a **Level** in the project. If the level is adjusted vertically, the floor will move with it. The associated level selection can be changed in the floor's properties but is initially set to match the associated level of the view the floor is placed in; e.g. placing the floor in the Level 1 floor plan view will associate the floor with level 1.

quick steps

Floor: Architectural

1. Ribbon [in sketch mode]
   - Draw *Panel*
     - Sketch enclosed boundary or add slope arrow.
     - Segment or Pick Options
     - Span Direction
   - Work Plane *Panel*
     - Set/Show Work Plane
     - Create Reference Plane
     - Viewer
2. Options Bar (only for Sketch Ceiling)
   - Offset
   - Extend into wall (to core)
3. Type Selector
   - Select Floor Type
4. Properties
   - Level & Offset
   - Room Bounding
   - See Warning
5. In-Canvas
   - First segment defines *Span Direction* by default
   - FYI: Top of floor assembly aligns with level
6. Green Checkmark to Finish Sketch
   - Objective is to sketch enclosed outline in a plan or 3D view

The top of a floor aligns with a level by default. Also, a floor gets thicker from the top down; i.e. the top edge does not move unless the Offset value is adjusted.

Know these basic settings while creating a floor in Revit:

1. **Pick Walls**: *selecting a wall defines an edge of the floor*
2. **Extend into wall (to core)**: *the floor extends past the wall finishes to its structure*
3. **Type Selector**: *what is the floor construction?*
4. **Level and Offset**: *where is the floor vertically within the project?*

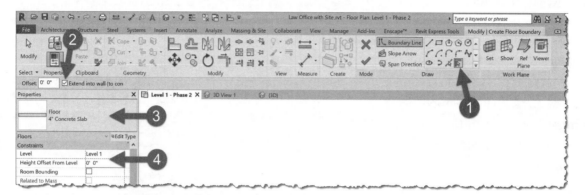

If the perimeter needs to deviate from the walls, switch from Pick Walls to Line, Arc, etc.

When a floor is selected, several properties are available via the Properties Palette, such as:

- **Perimeter**: *how much formwork is needed to pour concrete?*
- **Area**: *how much carpet is needed to cover this floor?*
- **Volume**: *how much concrete is needed to pour this slab?*
- **Thickness**: *how strong is this floor?*

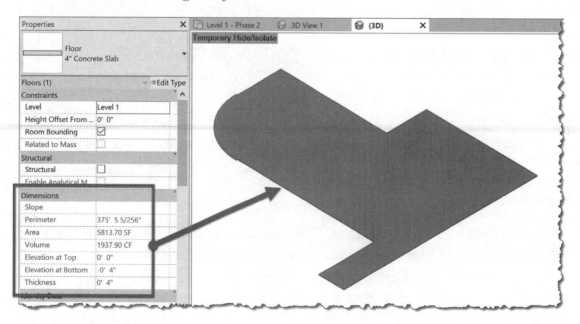

Modify elements using Align, Offset, Mirror, and Split tools

Modifying previously created content in Revit is an important task, and often performed more than time spent creating elements. Thus, the certification exam has several questions related to modifying elements within Revit. Here we will take a look at align, offset, mirror, and Split tools.

Align tool

The **Align** tool can be used to move one element to align with another. This works on most elements within Revit. For example, one wall can be moved to be in alignment with another wall. A table can be aligned with another table. Chairs can be aligned with chairs on the other side of a table. But this also works between element types as well. For example, a table can be aligned with the edge of an adjacent wall, or a chair with the centerline of a structural column.

The **Align** tool is activated via the Modify tab on the Ribbon.

You can use the ALIGN tool on the Ribbon to accurately move walls in alignment with other walls. The steps are the same as in the previous chapters!

Follow these steps to align two elements:
- *Select the reference (i.e. the one that will NOT move)*
- *Select the element to move*
- *Optional: click the padlock icon to lock the alignment*

When an alignment has been locked, moving either of the walls will cause the other to move as well. To break this constraint, select one of the elements and then click the padlock again, to unlock it.

An example of the Align tool being used to align two walls can be seen on the next page.

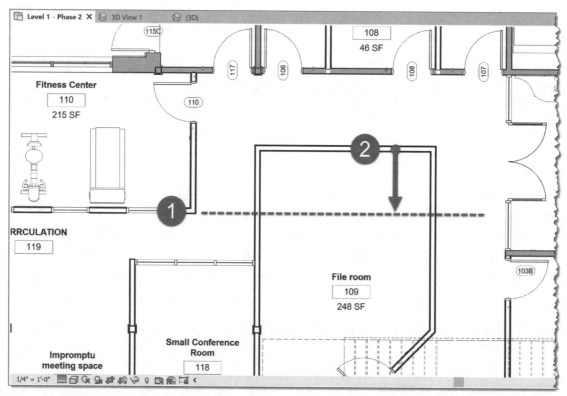

Using the Align command; pick the reference first, and then the element to move second

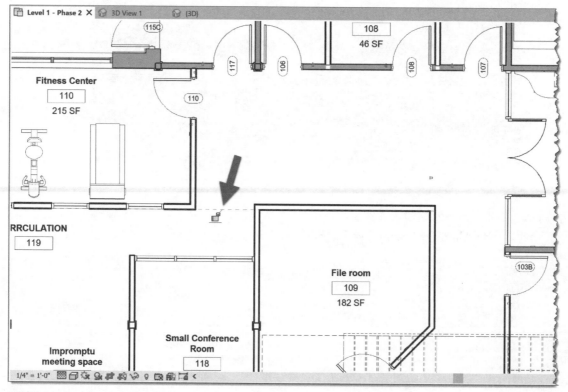

Wall moved using Align command; lock alignment opportunity

Offset

Use the Offset command to copy an element, such as a wall, beam, line, etc.) a specified distant, perpendicular to the side specified. For example, offset a wall to define a corridor.

Here are the steps required to offset an element:

- Start Offset command from Modify tab
- Options Bar
  - Graphical, define offset distance by picking points on-screen
  - Numerical, enter an offset value
- Select an element to offset
- Pick which side to offset towards

When using the offset command, the dashed line appears adjacent to the element to be offset, to help visualize the distance and direction before making the final click with the mouse.

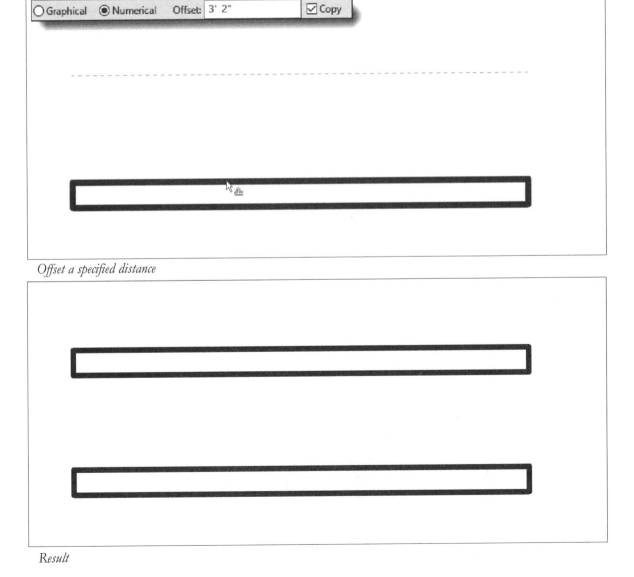

Offset a specified distance

Result

Mirror

Use the Mirror command to make a mirrored copy of selected elements.

 If you have an element, such as a line or wall, that represents the axis of reflection, use the **Mirror - Pick Axis** command. Its steps are outlined to the right.

 If you do not have an element, such as a line or wall, that represents the axis of reflection, use the **Mirror - Draw Axis** command. Its steps are outlined to the right. An example of the steps required are shown on the next page.

Using the **Mirror – Pick Axis**, Revit wants you to select a previously drawn line (or wall) to be used as the "axis of reflection," but if you do not have a line that would allow you to mirror the door properly, you have to pick the **Mirror - Draw Axis** option on the *Ribbon*.

1. In the example shown below, select the rectangle and the arc, use a *crossing window* by picking from right to left, and then pick the **Modify | Lines → Modify → Mirror - Draw Axis** tool from the *Ribbon*.

2. Make sure *Copy* is checked on the *Options Bar*. See image to the right; this is the default.

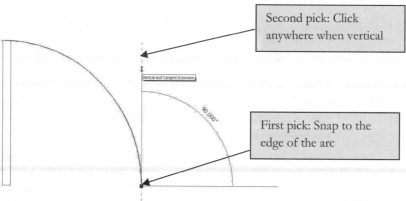

Mirror element using Draw Axis

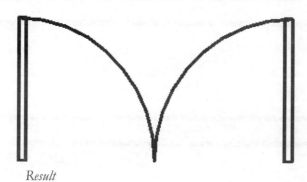

Result

Split

The Split tool has two options, one divides a wall into two elements with no gap, while the other also results in separate walls but inserts a fixed gap.

Here are how these two commands work:

Split Element

After starting this command, simply click on an element (in plan, elevation, or section view) to split it. See the example on the left side of the image below.

Split with Gap

After starting this command, specify the gap distance on the Options Bar, then click on an element (in plan, elevation, or section view) to split it. See the example on the right side of the image below.

Both options result in separate elements which are aligned and locked. Selecting any of the resultant elements may be unlocked or deleted.

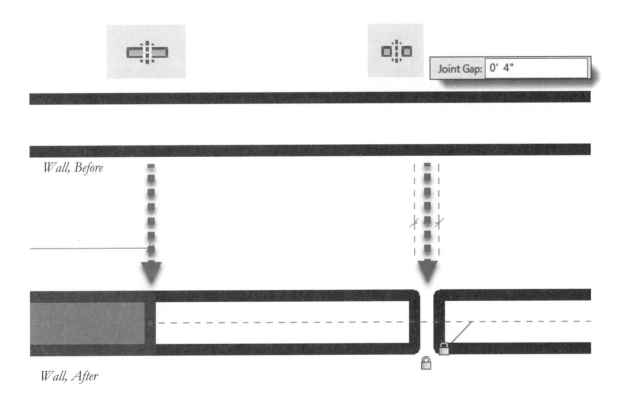

Move Modify Elements using Move, Copy, Rotate, Trim, and Extend

As previously stated, modifying previously created content in Revit is an important task. Thus, the certification exam has several questions related to modifying elements within Revit. Here we will take a look at move, copy, rotate, trim and extend tools.

Copy

The **Copy** tool allows you to accurately duplicate an element(s). You select the items you want to copy and then pick two points that represent an imaginary vector, which provides both length and angle, defining the path used to copy the object. You can also type in the length if there are no convenient points to pick in the drawing.

Move

The **Move** tool works exactly like the *Copy* tool except, of course, you move the element(s) rather than copy it.

You can use the MOVE tool on the Ribbon to accurately move walls.

Follow these steps to move an element:
- *Select the wall*
- *Click the Move icon*
- *Pick any point on the wall*
- *Start the mouse in the correct direction; do not click!*
- *Start typing the distance you want to move the wall and press Enter.*

Rotate

The **Rotate** tool is used to adjust element orientation. Here is how this tool works:

- Select objects to rotate
- Select the Rotate command (from the Modify tab on the Ribbon)
- Options Bar
  - o Disjoin (optional): used to break constraints to adjacent elements
  - o Copy (optional): used to duplicate rather than "move" elements
  - o Angle (optional): used to enter specific rotation value
  - o Place (optional): used to relocate default center of rotation
- On-Screen
  - o Pick first point to define start of rotation
  - o Next, either
    - ▪ Pick second point to graphically define angle
    - ▪ Or, type rotation angle and press Enter

Selecting the correct **Center of Rotation** (Base Point).

You need to select the appropriate Center of Rotation, or Base Point, for both the Scale and Rotate commands to get the results desired. A few examples are shown in the image below. The dashed line indicates the original position of the entity being modified. The black dot indicates the base point selected.

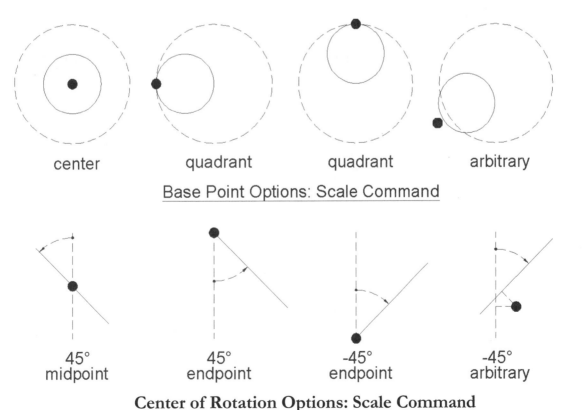

center   quadrant   quadrant   arbitrary

Base Point Options: Scale Command

45°   45°   -45°   -45°
midpoint   endpoint   endpoint   arbitrary

Center of Rotation Options: Scale Command

Trim and extend elements

Use the three **Trim/Extend** commands to edit lines, walls, ducts, etc. in Revit.

In addition to modifying existing elements, the use of the Trim/Extend tools typically results in a new constraint. Meaning, if one of the two modified walls/lines is edited, the other line will also adjust accordingly.

> To remove a constraint, using the **Move** command, with **Disjoin** selected on the *Options Bar*.

Here are the three Trim/Extend commands on the Modify tab, on the Ribbon.

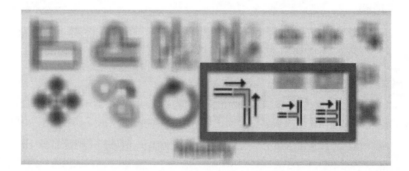

The next page highlights how these commands can be used. These examples are using simple lines. However, these steps also work on walls, ducts, pipes and lines in sketch mode (for floors, roofs, etc.).

Before…

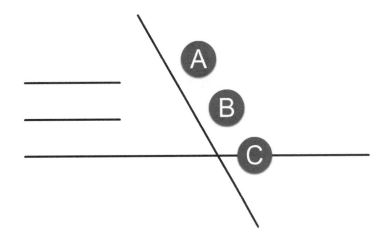

After…

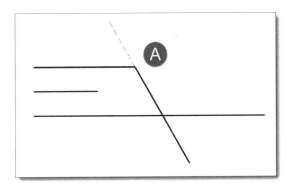

A. Trim corner:

- Trim/Extend to Corner
- Pick each line

TIP: Pick on the side to keep

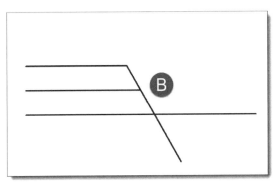

B. Extend line:

- Trim/Extend single element
- Pick the line to extend to
- Select line to extend

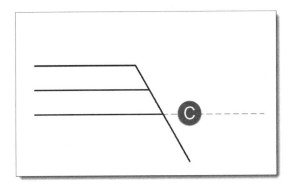

C. Trim line:

- Trim/Extend single element
- Pick the line to trim to
- Select the line to trim

TIP: Pick on the side to keep

Create & Modify Toposurface

The Toposurface command creates a 3D surface by picking points (specifying the elevation of each point picked) or by using linework within a linked AutoCAD drawing, that were created at the proper levels.

> Once a Revit surface has been created from a linked CAD file, there is no connection between the two elements. The Revit surface will not automatically update if the CAD file linked is modified.

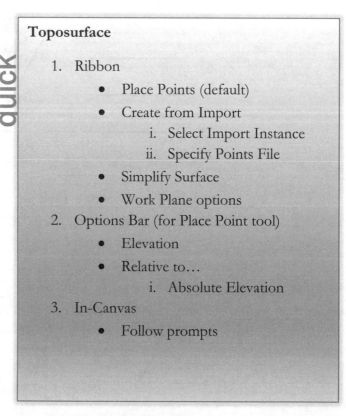

Toposurface

1. Ribbon
   - Place Points (default)
   - Create from Import
        i.   Select Import Instance
        ii. Specify Points File
   - Simplify Surface
   - Work Plane options
2. Options Bar (for Place Point tool)
   - Elevation
   - Relative to...
        i.   Absolute Elevation
3. In-Canvas
   - Follow prompts

Toposurface: Ribbon

Use **Place Points**, in conjunction with the current Elevation setting on the Ribbon, to manually define, or refine, a ground surface. This option is usually used in a site plan view.

Create from Import has two options: **Select Import Instance** and **Specify Points File**.

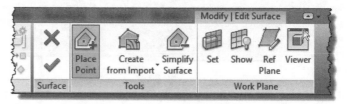

The first option is used against a linked in CAD file as discussed above. The second option is used if a "points file" is provided from a civil engineer or surveyor. This might come directly from a surveyor's data collector or exported from another application. However, the points should be reviewed and corrected, if needed, by a surveyor or civil designer before being used in Revit.

Simplify Surface will reduce the number of points while trying to maintain the integrity of the original surface. When a surface is created from a CAD link, the number of points in Revit can be quite large for large sites or highly detailed triangulated surfaces. This can have an impact on performance, so reducing the number of points can help. However, the surface needs to be closely inspected to verify its integrity.

The **Work Plane** panel has the usual **Set** and **Show** options. The **Ref Plane** command is offered for convenience and the **Viewer** command opens a temporary, throw-away, 3D view.

Toposurface: **Options Bar**

The **Elevation** controls the vertical position of any new, or selected, points. This option can be used when creating a surface for early phase conceptual surfaces. Another use is to manually follow the contours of a scanned drawing, which has been imported into Revit.

The drop-down controls what the elevation is relative to. The default option is **Absolute Elevation**, which corresponds to the **Project Base Point**.

Toposurface: **In-Canvas**

Working in a dedicated site plan view can be helpful. Looking at the View Range settings you can see the site is being viewed from 200′ above the first floor level, so your building/roof would have to be taller than that before it would be "cut" like a floor plan. The View Depth could be a problem here: on a steep site, the entire site will be seen if part of it passes through the specified View Range. However, items completely below the View Range will not be visible. In this case, the **Bottom** and **View Depth** can both be set to Unlimited to avoid any visibility issues.

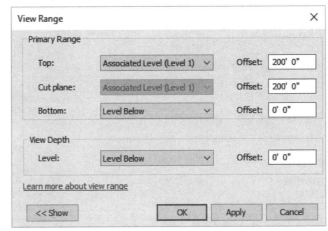

Figure 8.3-1 View Range settings for site plan view

A surface must have at least three points. Click the **Modify** button (not the tab) and select previously placed points to change their **Elevation** on the Options Bar.

Click the green checkmark to finish and the red "x" to cancel the current toposurface.

Various elements in the toposurface can be graphically controlled in **Object Styles** (project wide settings) and **Visibility/Graphic Override Settings** (view specific settings).

Create & Modify Columns

An Architectural Column is meant to be a non-load bearing element, often representing a finish around a structural column (e.g. studs and gypsum board).

These columns do not have a Structure, i.e., layers of materials, that can be edited as walls, ceilings and floors. Rather, they take on the structure of any wall they come in contact with.

Notice in Figure 3.1-9 that the Architectural Column at the top is a simple outline. The second, or middle column, is the same family, but has taken on the properties of the stud wall that runs through it.

> Using the **Join** command on the isolated column and the adjacent stud wall would apply the wall's properties even though they do not touch.

quick steps

Column: Architectural

1. Ribbon
   - Load Family
   - Model In-Place
2. Options Bar
   - Rotate after Placement
   - Level (3D Views)
   - *Select Direction*
     - Depth (downward)
     - Height (upward)
   - *Select Top Constraint*
     - Pick a Level *or*
     - Unconnected
       - i. Enter Height
3. Type Selector
   - Select Column Type
4. Properties
   - Moves with Grid
5. In-Canvas
   - Spacebar to rotate

Architectural Column: Ribbon

The Load Family and Model In-Place commands are located here for convenience.

Architectural Column: Options Bar

Checking the **Rotate after placement** option will immediately start the Rotate tool after placement. Checking Room Bounding limits the perimeter of a Room or Space element within a given room.

When is **Height** versus **Depth** used? See the previous comments on Walls and Structural walls in this chapter. Selecting Unconnected, in the second drop-down, creates a non-parametric column whose height will not change.

Architectural Column: **Type Selector**

The most used Architectural Column families are **Rectangular Column** and **Round Column**. The Column folder has a few other options.

Keep in mind that the Architectural Column tool existed prior to the Structural Column tool—Revit Architecture came before Revit Structure. One might notice the Column folder has a **Wood Timber Column** family and the Structural Column folder has a **Timber-Column** family. The version in the structural folder should be used. The architectural version will take on properties of any walls that touch it and, as previously stated, are meant to be non-bearing.

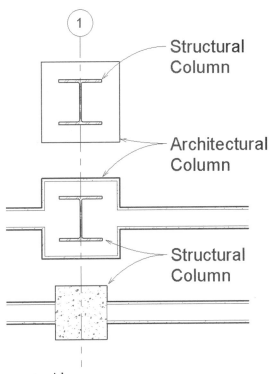

Architectural Column: **Properties**

Checking the **Moves with Grid** option will cause the column to maintain its relative position to the nearest grid.

If the **Material** Type Property is changed, from By Category to a specific material, the column will stop taking on the adjacent wall's material— the structure (aka layers) will still appear.

Architectural Column: **In-Canvas**

Tapping the **spacebar** rotates the column 90 degrees. When near an angled wall, and while tapping the Spacebar, the column will align with the wall.

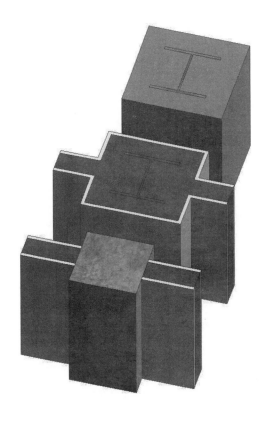

Notes:

6.0 Managing Views

Change the view scale

The **View Scale** controls the size of the text and symbols as well as the size of the drawing when placed on a sheet. For example, when the View Scale is set to **1/4" = 1'-0"** notice the size of the room tag. Now, compare that with the next image where the View Scale has been changed to **1/8" = 1'-0"**. This setting also affects the look for drafting patterns, like concrete or CMU shown in section.

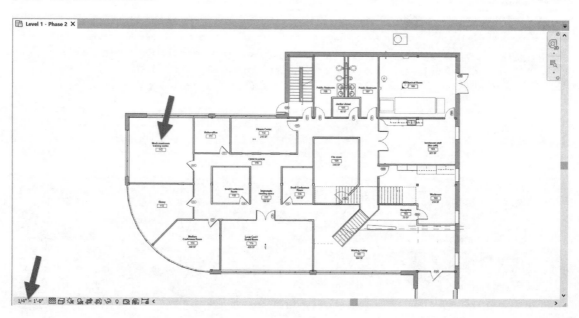

Floor plan with view scale set to 1/4" = 1'-0"

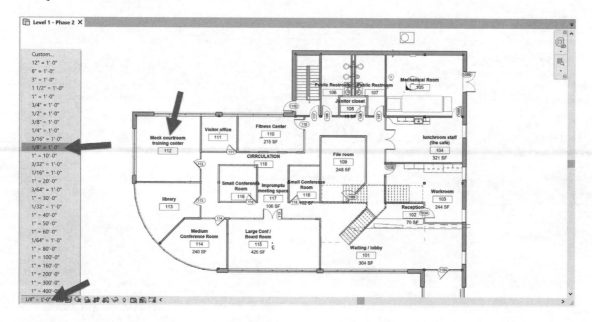

Floor plan with view scale set to 1/8" = 1'-0"

Change the detail level of a view

Each view has a **Detail Level** setting with three options: **Coarse**, **Medium** and **Fine**. When set to *Coarse*, walls appear as outlines and some content is shown simplified. When set to Fine, the walls show layers with fill patterns and some content appears more detailed. For example, in the two images below, the casework does not show the hardware or sorting shelves when the view is set to *Coarse*. However, they do show when Detail Level is *Fine*.

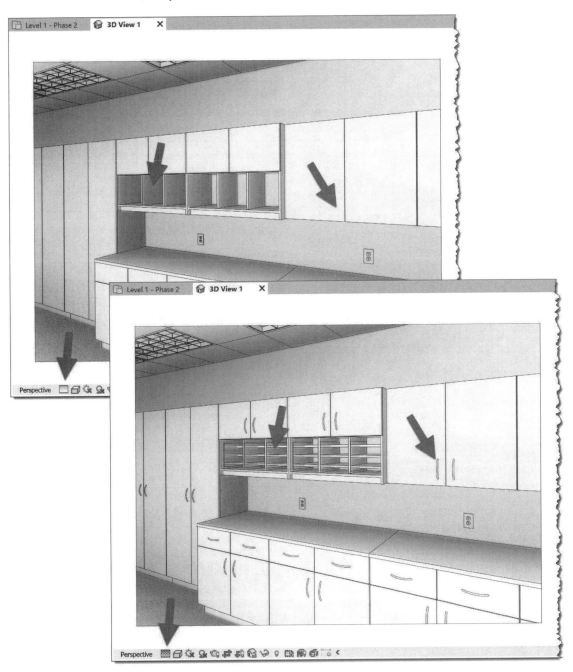

Manage visibility/graphics overrides for model categories

The Visibility/Graphics command opens the **Visibility/Graphics Overrides** dialog—which controls many related settings for a single view. This dialog is like 'Grand Central Station' when it comes to managing what and how things are seen in a given view.

The quicker way to get to this dialog is to type **VV** or **VG** on the keyboard.

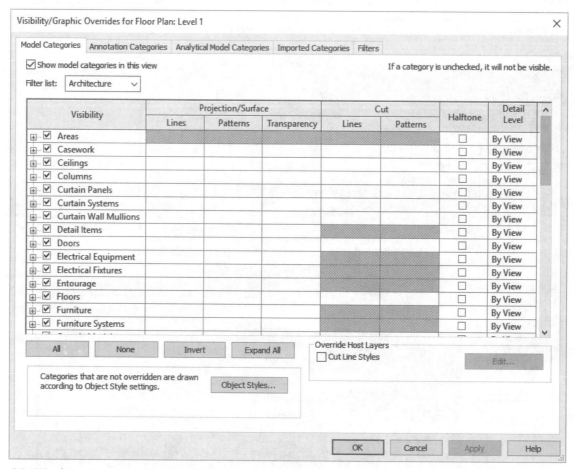

Visibility/Graphics Overrides dialog; Model Categories tab

The **Model Categories** tab contains the settings to control model elements in the current view—that is, things you can put your hands on when the building is actually built.

> Tags automatically disappear when its corresponding model category is turned off.

The **Annotation Categories** tab contains visibility control for elements used to annotate the model, typically for printed documents.

Many categories can have their line or pattern overridden or be set to transparent. Any grayed-out cells cannot have an override applied. For example, Revit does not allow furniture to be cut.

This would look odd in plan views for sure. Anytime an element is touched by a cut plane, in plan, elevation or section, the entire element appears.

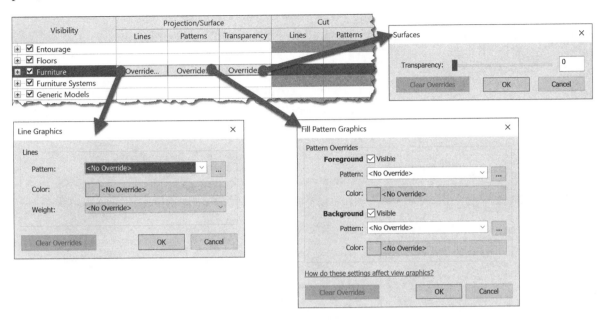

Visibility/Graphics Overrides dialog; Model Categories tab; Overrides

If the color of a category has been changed, for example, open the Visibility/Graphics Overrides dialog and click on the cell with the override to see what the changed color is, or to clear the override using the button in the lower left as seen in the images above.

All of these settings can be managed globally by selecting a **View Template** for each view type, for example, setting all floor plan views to **Architectural Plan** as shown in the following image.

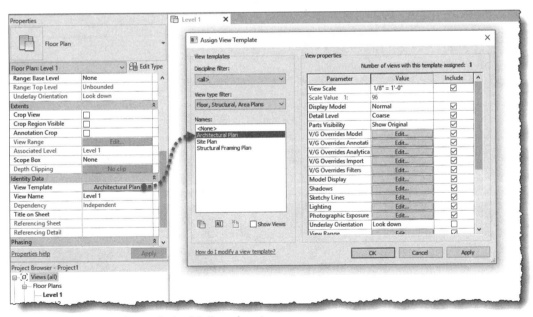

Assigning a view template to the level 1 floor plan view

Temporarily hide/isolate elements and components

The Temporary Hide/Isolate Elements tool is used to modify visibility without accidentally forgetting and unintentionally changing a view. When a view contains temporarily hidden/isolated elements, its edge is highlighted with a cyan colored border.

Simply select element to hide/isolate and pick an option from the display control bar as shown in the image below. When done, use the **Reset Temporary Hide/Isolate**, or **Apply** to make the changes permanent in this current view. Close/re-open the model also resets the view.

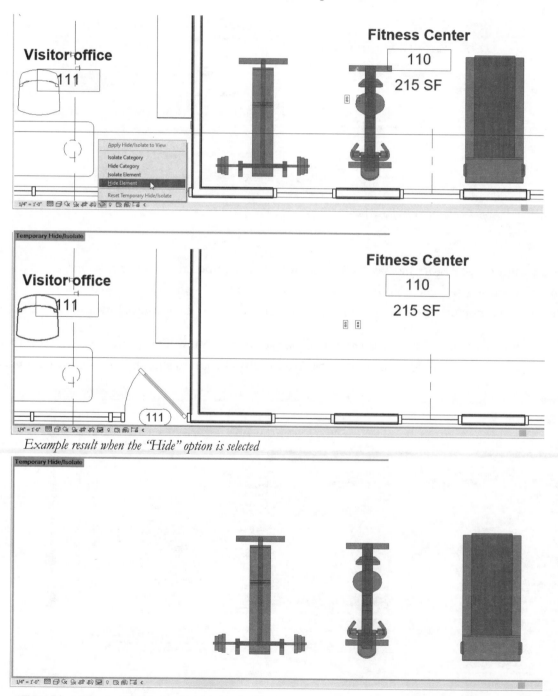

Example result when the "Hide" option is selected

Example result when the "Isolate" option is selected

Manage view range

Each floor plan has its own **View Range** settings, which are accessed via the Properties Palette (when nothing is selected in the model and a plan view is current).

For a plan view, architecturally, not much appears above the cut plane. The main setting here is **Cut Plane**. This determines if windows appear and where stairs are cut for example.

Example: In the image to the right, the Cut Plane is 4'-0".

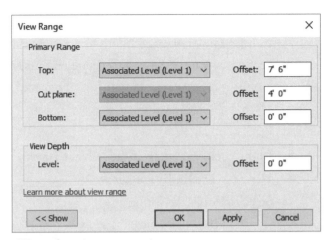

Floor plan view range settings

The **Bottom** setting is where Revit stops looking. Thus, even when a floor element has not been created yet, the elements on the floor/level below will not appear.

The **View Depth** section allows some elements, which fall between the **Bottom** and **View Depth**, to be overridden with a special **Line Style** called **<beyond>**. The View Depth must be equal to, or lower than, the Bottom setting.

For RCP (reflected ceiling plan) views, the **View Range** options are a little different. Click the "Learn more about view range" link to access more information on this topic. Also, click the **<<Show** button to see a diagram showing how this feature affects the view.

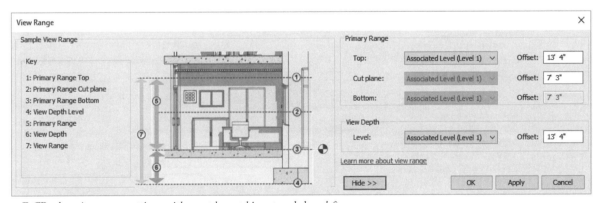

RCP plan view range settings with sample graphic expanded on left

Duplicate views

Views are often duplicated to separate information into clear documentation. For example, a floor plan might be duplicated multiple times to show dimensions, fire ratings, interior finishes, and more. Each view can have different elements/categories shown or hidden as well as unique annotation elements, such as dimensions, tags, and text.

Steps to duplicate a view:

- Right-click on a view in the Project Browser
- Hover over **Duplicate View**
- Select a duplicate option:
  - **Duplicate**: Just the view is duplicated (includes view properties)
  - **Duplicate with Detailing**: Includes copy of annotation elements
  - **Duplicate as a Dependent View**: Tied to original view, with separate crop area
- Rename new view (optional)

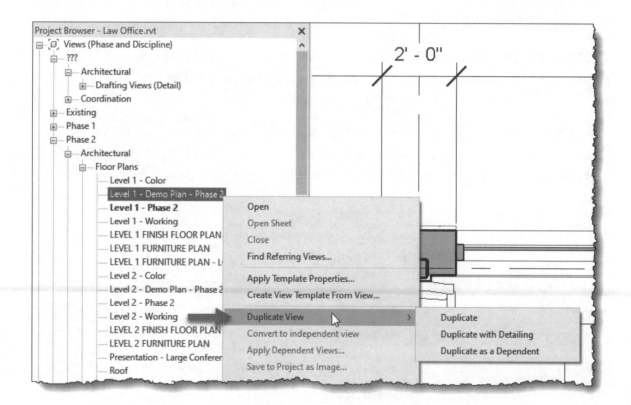

A view can only be placed on a single sheet in Revit. Thus, views may also need to be duplicated if the same view is required on more than one sheet – which can also be at a different scale if needed.

Create section views

Section and elevations are essentially defined by vertical cuts through the model. A section will show walls/floors/ceilings/roofs being cut, while an elevation is cropped down to be within the boundaries of those same elements. This section covers the creation of section views.

<u>Section</u>

Use this command to create a section view of the Revit model—the result is a graphic in the model representing the cut plane and view direction and a new item in the Project Browser. Typically created in a floor plan view, a section view is created by simply clicking two points. The direction picked—left to right, or right to left—determines the view direction. However, the view direction can be changed later if needed. The image below lists some terms related to the *Section* graphic.

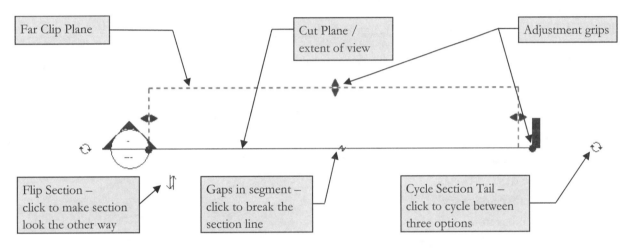

Selecting a section shows the far clip plane as shown below.

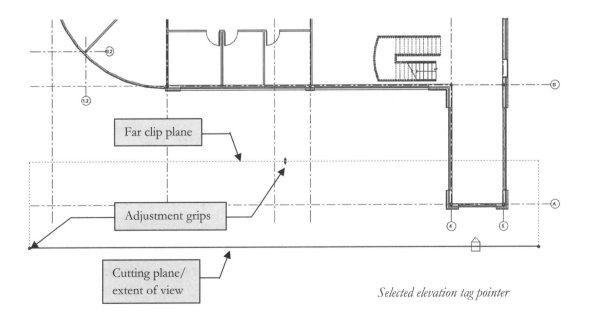

Selected elevation tag pointer

Create Elevation views

Section and elevations are essentially defined by vertical cuts through the model. A section will show walls/floors/ceilings/roofs being cut, while an elevation is cropped down to be within the boundaries of those same elements. This section covers the creation of elevation views.

Elevation

This command is used to create exterior or interior elevations.

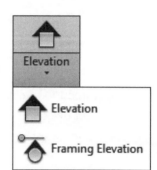

From a floor plan view, click near a wall to be elevated – before clicking, be sure the pointer on the elevation tag is properly positioned. The pointer automatically changes, during placement, to point at the nearest perpendicular wall.

The **Project Browser** now has a new view listed in the elevations section. Double-click to open or right click to rename or delete it.

Once placed in a plan view, selecting the center portion of the elevation tag causes a few graphic options to appear as shown below.

When the "body" of the elevation tag is selected, there is an option to "turn on" the other elevations by checking the desired boxes. Doing so will cause additional elevations to appear in the Project Browser. However, selecting and deleting the "body" part will cause all related elevations to be deleted.

Rotation Control icon visible when Elevation Tag is selected

Unchecking a box will also delete a view from the Project Browser.

Click here

Selected elevation tag

In each Section or Elevation view, use the **Visibility/Graphics Overrides** dialog (type VV or VG to open) to control visibility of elements and how they look (line weight, color, etc.). Using a **View Template** will apply consistent settings to all views of the same type (e.g. all interior elevation views). A View Template is applied via the Properties Palette for each view.

Create 3D views and Renderings

3D views aid in the creation and validation of a Revit model. They are also the basis of photo realistic renderings often used to present the design solution to the client.

Create a 3D View

Clicking the **Default 3D View** command opens a 3D view named **{3D}** and appears under the **3D Views** heading in the **Project Browser**. This is a quick way to see an exterior overview of the project. If this view is ever renamed or deleted, a new one is automatically created the next time the **Default 3D View** command is selected.

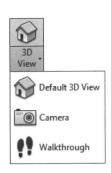
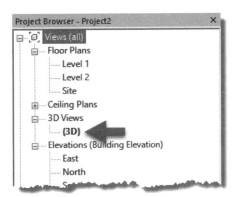

A **Camera** view is a perspective view, which looks more realistic than the Isometric view (Default 3D View just covered).

Camera views are typically placed in a plan view. Here are the basic steps:

- Open a **floor plan** view
- Select the **Camera** command
- Notice **Offset** value on Options Bar
    - This is the "eye" height camera
- Pick a point to locate the **eye position**
- Notice **Offset** value on Options Bar again
    - This is a point at which the sightline will pass through
- Pick to define the view **target location**
- Optional: Turn off **Far Clip Active** so elements are not being omitted.

The **Eye** and **Target** Elevations can be modified at any time. Use the **Far Clip Offset** to control how far into the model you can see. The camera command may be accessed from the Quick Access Toolbar, View tab or keyboard shortcut.

The **Crop Region** can be turned off in perspective views (i.e. camera views). In an uncropped perspective, using the scroll-wheel on the mouse will move the camera within the scene, rather than just zooming (like in a cropped perspective).

Create a Rendering

Revit offers the ability to create a photo-realistic rendering from a Revit model. By adding lighting and properly applying materials, amazing results can be achieved.

To access the rendering dialog, while in a 3D view, click the **Show Rendering Dialog** icon on the View Control Bar in the lower left. This opens the **Rendering** dialog shown to the right. This dialog box allows you to control the environmental and quality settings and create the rendering.

After adjusting the quality, output, lighting, background, and image options, click the **Render** button.

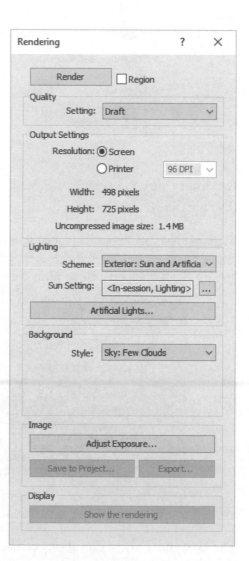

After a few minutes, depending on the speed of your computer, you should have a rendered image similar to the example shown below. You can increase the quality of the image by adjusting the quality and output settings in the Render dialog. Keep in mind these higher settings require substantially more time to generate the rendering.

When the rendering is complete, use either the **Save to Project** or **Export** buttons shown in the Rendering dialog above. The first option will add an item within the Project Browser, and the latter will create a separate jpg or png file on your computer.

7.0 Managing Documentation

Create and modify text

The first thing to know about the text tool is that it should be used as little as possible! Rather, live tags, keynotes and dimensions are preferred over static text to ensure the information presented is correct. Text will not update or move when something in the model changes—especially if the text is not visible or in the view where the model is being changed.

The Text tool can be started from the **Quick Access Toolbar**, the **Annotate** tab or by typing **TX** on the keyboard.

Text can be placed in any view and on sheets. The only exception is the Text command does not work in schedule views.

> When text is selected, it can be changed via the Type Selector in the Properties Palette.

To place text in a 3D view, the view must be Locked via the Save Orientation and Lock View command on the View Control Bar at the bottom. Also, use the Set Work Plane before placing text to control how text looks. It is not possible to have a leader with no text.

Text can also be in a **Group**. When the group only has elements from the Annotate tab, it is a Detail Group. When the group also has model elements, the text is in something called an Attached Detail Group. When a Model Group is placed, selecting it gives the option adding the Attached Detail Group.

When text is selected, click the **Edit Type** button in Properties to open the Type Properties shown to the right. Notice this is where the **Font Size** is defined (i.e. the height on a printed page) as well as the **Text Font**; Arial is one of the installed Windows fonts on the computer.

quick steps

Text

1. Ribbon
   - Various format options – see notes in this section
   - Check Spelling
   - Find/Replace
2. Type Selector
   - Select text type
3. In-Canvas
   - Single click and start typing or click and drag to define width and then start typing
   - Enter create a new line
   - Click away from text to finish the text tool

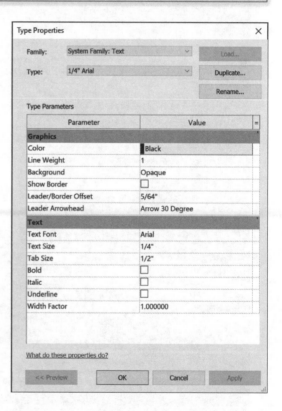

Create and modify dimensions

All the commands available for dimensioning the model can be found here. This image also shows the **expanded panel area** which provides access to modify and create (via Duplicate) types for each dimension command.

Aligned dimension tool is the most used of the multiple dimensioning tools. Its purpose is to graphically indicate the distance between two elements or points. The Aligned dimension is dependent on two picked elements or points. Like a door or window being dependent on a wall to exist, if a referenced element is deleted, the dimension is deleted as well. Keep in mind, the deleted dimension may not be in the same view where the model element is being deleted.

The resulting dimension is based on how the two reference picks are made. In the image below, the "Aligned" dimension on the right was created by selecting directly on the angled line on the <u>left</u> and then tapping **Tab** to pick a point on the <u>right</u>. The other "aligned" dimension used **Tab** to pick points for <u>both</u> picks.

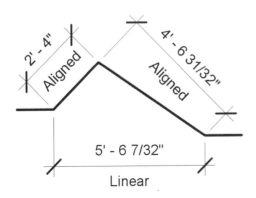

quick steps

Dimension: Aligned

1. Ribbon
   - Option to switch/toggle to another dimension tool
2. Options Bar
   - 'Dimension to' setting
     i. Wall Centerlines
     ii. Wall Faces
     iii. Center of Core
     iv. Faces of Core
   - Pick
     i. Individual References
     ii. Entire Walls
   - Options button
     i. Available when Pick: Entire Walls is selected
     ii. Select what to automatically dimension to on a selected wall
3. Type Selector
   - Select dimension type
4. Type Properties
   - Units Format; rounding
5. In-Canvas
   - Tap the Tab key to toggle between layers in a wall
   - Picking multiple elements will create a dimension string
   - Picking same element a second time will remove the reference
   - Click the padlock icon after placing a dimension to lock it
   - Click the "EQ" icon after placing a dimension string to equally space the dimension elements

For the dimension on the right, notice the witness line is parallel (or aligned, as the name implies) with the selected element. By contrast, the **Linear** dimension is always horizontal or vertical; this command is covered next.

For all dimension commands, the length value can be modified by clicking directly on the text—which opens the Dimension Text dialog shown to the right.

> When deleting a locked dimension, you are prompted to keep or remove the constraint.

Here, text can be added as a prefix, suffix, above and below the dimension value. To change the dimension value itself, click the **Replace with Text** option. Revit prevents faking dimensions by entering another length value.

When a dimension is selected, some allow it to be locked so the length value cannot be changed. This will cause other geometry to move rather than allowing the dimension to change.

To undo a dimension override, switch back to **Use Actual Value** in the Dimension Text dialog.

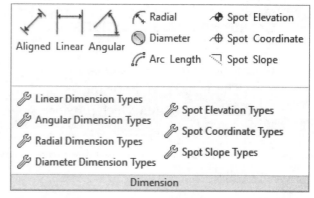

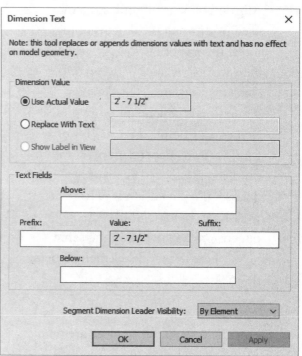

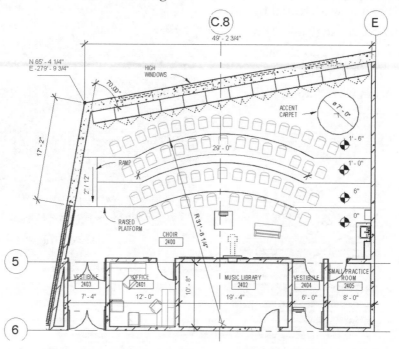

Create and modify a sheet

Revit views, legends and schedules are composed on sheets in Revit, to create printed drawings for presentation and construction documents (CDs).

To create a sheet: **View → Sheet Composition → Sheet**. This opens the New Sheet dialog shown to the right. Pick from an available titleblock and click ok. A new Sheet now appears in the Project Browser and can be renamed and renumbered.

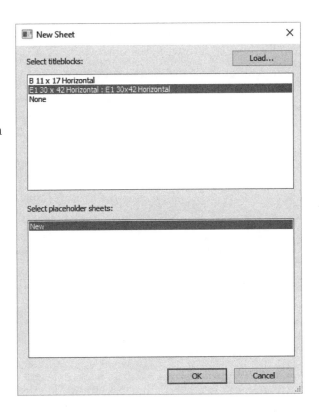

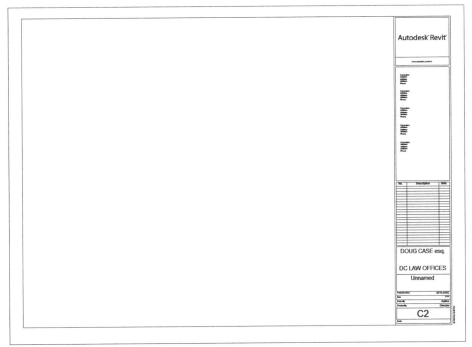

Place plan views on a sheet

To place a view on a sheet, first open the sheet and then drag and drop a view from the Project Browser as shown in the image below. A view can only be placed on one sheet in order to maintain the drawing and sheet references automatically.

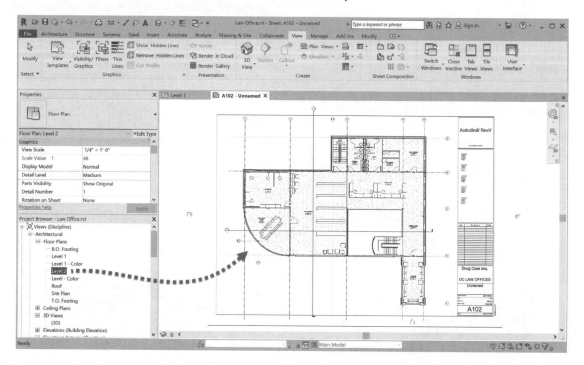

Revit automatically assigns a **Detail Number** to each view placed on a sheet. This number can be modified in that view's properties but cannot be the same number as another view on the same sheet. The drawing title matches the view name in the Project Browser unless the view's **Title on Sheet** parameter is used. The scale of the view is based on the View Scale. If the View Scale is changed, the view will automatically update on the sheet – which will likely require the view to be manually repositioned on the sheet.

In the titleblock, some parameters are project wide while others are per sheet. The project number, for example, is automatically applied to all sheets when entered once. By contrast, the Drawn By field is a per sheet setting as projects can have any number of people working on them.

Create and modify schedules

Schedules are a powerful feature in Revit, used to quantify modeled elements and view/organize their properties.

When starting the command, from **View →Schedules → Schedule/Quantities**, the first step is to select the **Category**. The **Name** and **Phase** can be changed later, but the Category option is a one-time selection; if wrong the schedule needs to be deleted and recreated.

Every element in Revit goes in a specific Category, which is defined when the content was initially created.

The list of Categories is predefined by Revit and cannot be changed.

Most of the *Category* names are straightforward; walls go in the *Walls* category, doors in the *Doors* category, and so on. Below are a few examples that might not be as easy to deduce.

- **Casework**: Base cabinets and wall cabinets, reception desk, built-in bookshelves
- **Furniture Systems**: Cubicles and other built-in office furniture
- **Specialty Equipment**: Toilet room accessories, lockers, ladders, etc.
- **Topography**: Site or ground surface

When a Category is selected for a new schedule, Revit knows to look for all the elements that have been placed in the model which are in that Category. A **<Multi-Category>** schedule has the ability to look at every element in the model and report on parameters they all have in common, such as Cost.

Schedule Properties

Once a schedule is created, you are immediately brought to the **Schedule Properties** dialog (shown to the right). This is the place you select what information you want listed in the schedule and what it should look like.

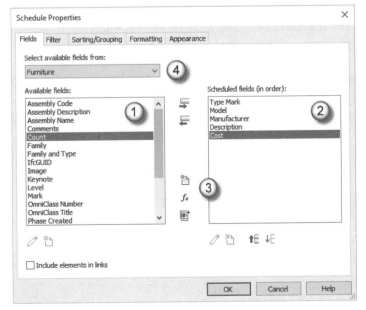

You will notice the Schedule Properties dialog is divided into five separate areas; each area has its own tab near the top.

Before Revit will create the schedule, you must select one or more parameters to be scheduled. To do this you add **Available fields**, listed on the left, to the **Scheduled fields (in order)** list, on the right. Simply select a parameter on the left and click the **Add →** button, or just double-click the parameter name.

If you simply selected the five parameters shown on the right (Figure 10.3-32, #2) and clicked OK the schedule would be created. A new item would appear in the Project Browser, under **Schedules/Quantities**. If you are in a new or empty project, the schedule would be empty (Figure 10.3-34). On the other hand, if the model has 30 items in the selected category, you should see 30 lines/rows, one for each instance in the model. Additionally, if those 30 elements have information entered into any of the scheduled parameters, this information will appear in the schedule automatically.

<Furniture Schedule>				
A	B	C	D	E
Type Mark	Model	Manufacturer	Description	Cost

While in a schedule view, the Properties Palette shows information related to the current schedule (see image to right). Clicking any of the **Edit** buttons will open the **Schedule Properties** dialog, allowing further refinement to the schedule).

Schedules are listed in the Project Browser, and can be renamed, duplicated, or deleted.

Schedules may be placed on sheets, just like other views. However, some schedules are just meant to organize data and are never placed on sheets. But schedules cannot be printed unless placed on a sheet. Finally, schedules can be exported to a comma-delimited text file, which can then be opened in a spreadsheet application, such as Microsoft Excel.

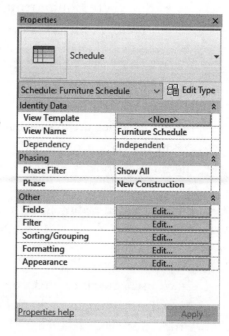

This section is provided to help you test your Revit "architecture" knowledge before taking the provided practice exam or the certification exam. The answers to the questions can be in the next chapter. Keep in mind these questions are focused on the architectural "user" exam, not professional, MEP or structural exams.

> This practice test can also be taken using the provided electronic version. See the inside-front cover for instructions and your unique access code.

The questions are made up by the author and should not be memorized, thinking the same questions might be asked on the exam. These questions are similar in that they relate to the posted exam objectives on the Autodesk and Certiport websites.

It is recommended that you wait to take this sample test until you have reviewed all the previous chapters in this book.

Directions: Answer the following questions to test your Revit knowledge.

TIP: The software helps you answer the question in the correct way. That is, if you are expected to select only one answer you can only select one; this looks like a circle. If you are expected to select all that apply, you select one or more square boxes:

        O = Select only one answer

        □ = Select all that apply

1. Which of the following is NOT a valid *Detail Level Style*?
   - o Rough
   - o Coarse
   - o Fine
   - o Medium

2. The Cut Plane setting is found in which dialog?
   - o Properties Palette
   - o Project Information
   - o View Range
   - o Type Properties

3. When a railing is selected, its total length is listed in the Properties Palette.
   - o True
   - o False

4. Which are TRUE about a titleblock in Revit?
   - ☐ "Sheet number" must be unique
   - ☐ Are found on sheets
   - ☐ Project title automatically appears on all sheets
   - ☐ "Drawn by" field must be unique

5. For Trim/Extend to Corner command, click the portion of wall/line to delete.
   - ○ True
   - ○ False

6. A *Room* has to be created in each level 1 plan you wish to see a room tag in. For example: Level 1 Floor Plan, Level 1 Ceiling Plan, Level 1 Finish Plan.
   - ○ True
   - ○ False

7. What is NOT possible with the *Text* tool?
   - ○ Add a leader with no text
   - ○ Center justify the text
   - ○ Change the style
   - ○ Add multiple leaders

8. Which are true about *Detail Views*?
   - ☐ The *View Scale* can be adjusted
   - ☐ They update geometry when the model updates
   - ☐ They are 2D drawings
   - ☐ A different text tool is needed to add notes, compared to a model view

9. While modeling a stair, you can:
   - ○ not adjust the stair width
   - ○ stop the sketch short of the specific level
   - ○ end up with various size risers
   - ○ not use snaps

10. Which are true about railings?
    - ○ Added automatically with stairs
    - ○ Can adjust height on options bar while sketching path
    - ○ Automatically added to edge of interior floor openings
    - ○ Style can be changed, when selected, via the *Type Selector*

11. Select the item (i.e. component) that can be loaded into the project from a file:
    o Wall
    o Door
    o Floor
    o Roof

12. Which THREE views are automatically created when a level is added to a project?
    ☐ Floor plan
    ☐ Structural Plan
    ☐ 3D view
    ☐ Ceiling plan

13. Double-clicking on a Ribbon tab can hide commands.
    o True
    o False

14. Adding an opening, e.g. door or window, changes a wall's area listed in properties.
    o True
    o False

15. Changing the View Scale does not affect the size of Room Tags in that same view.
    o True
    o False

16. Which are true statements about walls?
    ☐ Base Offset repositions the wall vertically
    ☐ During creation, Spacebar flips wall about location line
    ☐ Cannot be used to represent foundation walls
    ☐ Room Bounding by default

17. Select the statements that are true about doors.
    ☐ Doors may be placed in an elevation view
    ☐ Doors can only be deleted in plan views
    ☐ Doors are floor hosted
    ☐ Doors can be deleted from a schedule

18. A selected window can be accurately positioned using the temporary dimensions displayed.
    o True
    o False

19. Select the various constraints:
    - ☐ Equality
    - ☐ Pin
    - ☐ Shared Parameter
    - ☐ Locked Dimension

20. Which statement is true about camera views?
    - o The eye elevation cannot be adjusted
    - o The target elevation can be adjusted
    - o You cannot adjust how far into the model you see
    - o Camera views can only be created from the *Ribbon*

21. If you delete a locked dimension you have the option to keep the constraint.
    - o True
    - o False

22. Which key do you use to be able to select elements which are overlapped?
    - o Caps Lock
    - o Alt
    - o Tab
    - o F6

23. Which are window family *Type Parameters*?
    - ☐ Window Sill
    - ☐ Width
    - ☐ Height
    - ☐ Type Mark

24. Where are newly created views listed?
    - o Options Bar
    - o Application Menu
    - o Ribbon
    - o Project Browser

25. Pressing the Esc key, on the keyboard, unselects elements and cancels the current command.
    - o True
    - o False

26. Which are components of the User Interface?
- ☐ Options Bar
- ☐ Application Menu
- ☐ View Control Bar
- ☐ Project Browser
- ☐ Door Family

27. Which Statements are true about grid lines?
- ☐ Two grids can have the same number
- ☐ Show up automatically in elevation and section views
- ☐ Start and end points will automatically align if drawn correctly
- ☐ Grid heads display on either end or both

28. Which are ways to hide an element?
- ☐ Visibility/Graphics Override → Hide Category
- ☐ Right-click element → Hide Element
- ☐ Delete it
- ☐ Visibility/Graphics Override → View Filter

29. Element can only be arrayed in a rectangular pattern.
- o True
- o False

30. All floor edges must be aligned with a wall when created.
- o True
- o False

Answers:

See the next chapter for the answers to these questions.

ANSWERS:

1. Which of the following is NOT a valid *Detail Level?*
   - o Rough
   - o Coarse
   - o Fine
   - o Medium

2. The Cut Plane is found in which dialog?
   - o Properties Palette
   - o Project Information
   - o View Range
   - o Type Properties

3. When a railing is selected, its total length is listed in the Properties Palette.
   - o True
   - o False

4. Which are TRUE about a titleblock in Revit?
   - ☐ "Sheet number" must be unique
   - ☐ Are found on sheets
   - ☐ Project title automatically appears on all sheets
   - ☐ "Drawn by" field must be unique

5. For Trim/Extend to Corner command, click the portion of wall/line to delete.
   - o True
   - o False

6. A *Room* has to be created in each level 1 plan you wish to see a room tag in. For example: Level 1 Floor Plan, Level 1 Ceiling Plan, Level 1 Finish Plan.
   - o True
   - o False

7. What is NOT possible with the *Text* tool?
   - o Add a leader with no text
   - o Center justify the text
   - o Change the style
   - o Add multiple leaders

8. Which are true about *Detail Views?*
   - ☐ The View Scale can be adjusted
   - ☐ They update geometry when the model updates
   - ☐ They are 2D drawings
   - ☐ A different text tool is needed to add notes, compared to a model view

9. While modeling a stair, you can
   - o not adjust the stair width
   - o stop the sketch short of the specific level
   - o end up with various size risers
   - o not use snaps

10. Which are true about railings?
    - o Added automatically with stairs
    - o Can adjust height on options bar while sketching path
    - o Automatically added to edge of interior floor openings
    - o Style can be changed, when selected, via the *Type Selector*

11. Select the item (i.e. component) that can be loaded into the project from a file.
    - o Wall
    - o Door
    - o Floor
    - o Roof

12. Which THREE views are automatically created when a level is added to a project?
    - ☐ Floor plan
    - ☐ Structural Plan
    - ☐ 3D view
    - ☐ Ceiling plan

13. Double-clicking on a Ribbon tab can hide commands.
    - o True
    - o False

14. Adding an opening, e.g. door or window, changes a wall's area listed in properties.
    - o True
    - o False

15. Changing the View Scale does not affect the size of Room Tags in that same view.
    - o True
    - o False

16. Which are true statements about walls?
    - ☐ Base Offset repositions the wall vertically
    - ☐ During creation, Spacebar flips wall about location line
    - ☐ Cannot be used to represent foundation walls
    - ☐ Room Bounding by default

17. Select the statements that are true about doors.
    - ☐ Doors may be placed in an elevation view
    - ☐ Doors can only be deleted in plan views
    - ☐ Doors are floor hosted
    - ☐ Doors can be deleted from a schedule

18. A selected window can be accurately positioned using the temporary dimensions displayed.
    - o True
    - o False

19. Select the various constraints
    - ☐ Equality
    - ☐ Pin
    - ☐ Shared Parameter
    - ☐ Locked Dimension

20. Which statement is true about camera views?
    - o The eye elevation cannot be adjusted
    - o The target elevation can be adjusted
    - o You cannot adjust how far into the model you see
    - o Camera views can only be created from the *Ribbon*

21. If you delete a locked dimension you have the option to keep the constraint.
    - o True
    - o False

22. Which key do you use to be able to select elements which are overlapped?
    - o Caps Lock
    - o Alt
    - o Tab
    - o F6

23. Which are window family *Type Parameters*?
    - ☐ Window Sill
    - ☐ Width
    - ☐ Height
    - ☐ Type Mark

24. Where are newly created views listed?
    - o Options Bar
    - o Application Menu
    - o Ribbon
    - o Project Browser

25. Pressing the Esc key, on the keyboard, unselects elements and cancels the current command.
    - o True
    - o False

26. Which are components of the User Interface?
    - ☐ Options Bar
    - ☐ Application Menu
    - ☐ View Control Bar
    - ☐ Project Browser
    - ☐ Door Family

27. Which statements are true about grid lines?
    - ☐ Two grids can have the same number
    - ☐ Show up automatically in elevation and section views
    - ☐ Start and end points will automatically align if drawn correctly
    - ☐ Grid heads display on either end or both

28. Which are ways to hide an element?
    - ☐ Visibility/Graphics Override ☐Hide Category
    - ☐ Right-click element ☐Hide Element
    - ☐ Delete it
    - ☐ Visibility/Graphics Override ☐View Filter

29. Elements can only be array in a rectangular pattern.
    - o True
    - o False

30. All floor edges must align with a wall when created.
    - o True
    - o False

Notes:

Introduction

This chapter will highlight the practice exam software provided with this book, including accessing the exam, required files, user interface and how to interpret the results. Taking this practice exam, after studying this book, will help ensure a successful result when taking the actual Autodesk Certified User (ACU) exam at a test center.

The practice exam questions are similar, not identical, to the actual exam.

Important Things to Know

Here are a few big picture things you should keep in mind:

- **Practice Exam – First Steps**
  - The practice exam, that comes with this book, is taken on **your own computer**
  - You need to have **Revit installed** and ready to use during the practice exam
  - You must download the practice exam software from SDC Publications
    - See inside-front cover of this book for access instructions
  - **Required Revit files** for the practice exam
    - Files downloaded with practice exam software
    - Locate files before starting practice test
  - Note which questions you got wrong, and study those topics

- **Practice Exam - Details**
  - Questions: 30
  - Timed: 50 minutes
  - Passing: 70%
  - Results: Emailed (optional)

This practice exam can be taken multiple times. But it is recommended that you finish studying this book before taking the practice exam. The questions are presented in a random order, but there are only 30 questions total, so you don't want to get to a point where all the questions, and their answers, have been memorized. This will not help with the actual exam as they are not the same questions.

This practice exam can be taken multiple times.

Practice Exam Overview

The **practice exam** included with this book can be downloaded from the publisher's website using the **access code** found on the inside-front cover. This is a good way to check your skills prior to taking the official exam, as the intent is to offer similar types of questions in roughly the same format as the formal ACU exam. This practice exam is taken at home, work or school, on your own computer. You must have Revit installed to successfully answer the in-application questions.

This is a test drive for the exam process:

- Understanding the test software
- How to mark and return to questions
- Exam question format
- Live in-application steps
- How the results are presented at the exam conclusion

Here is a sample of what the practice exam looks like…

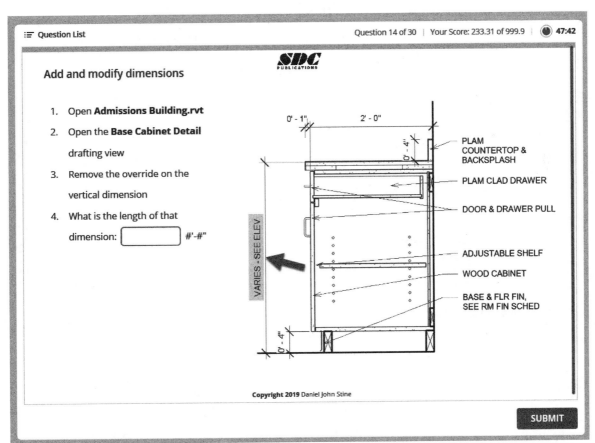

Sample question from included practice exam

If you enter your name and email address at the beginning of the practice exam, you will receive an email with your result. If you are required to take the practice exam for a class, the 'results email' can then be forwarded to your instructor. Alternatively, you could also enter your instructor email address. Whether you enter your name and email address or not, neither the author nor SDC Publications captures any data related to this practice exam.

Having taken the practice exam can remove some anxiety one may have going into an exam that may positively impact your career search.

Download the Practice Exam

Follow the instructions on the inside-front cover of this book, using the provided **access code** to **download** the practice exam. Once the ZIP file is downloaded you must extract the files into a folder that you create.

Suggested steps:

- Create a folder on your desktop or C drive, such as **C:\Autodesk Certified User – Practice Exam**
- Double-click on the downloaded ZIP file
- Copy all the folders/files from the ZIP file to the newly created folder

The image below shows the files copied to the recommended folder.

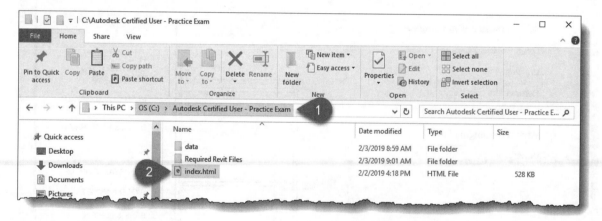

Practice exam files saved to a folder

Required Revit Files

As already mentioned several times, during the exam you must use Revit to complete a task and enter an answer based on the results created, such as the area of a wall or the distance between two items. For the practice exam, you must use the two provided Revit files: **Admissions Building.rvt** and **Lake Cabin.rvt** (see image below). Looking back at the sample question, a few pages back, you will notice the first step is to open the 'admissions building' Revit project file. Thus, it is important to know where these files are located on your computer before starting the timed exam.

For the practice exam, the files are in the same ZIP file as the exam itself.

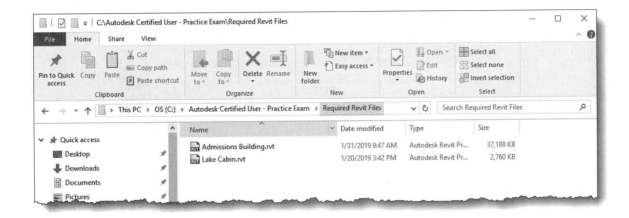

Provided Revit files required to complete practice exam

Starting the Practice Exam

To start the practice exam, double-click on the **index.html** file pointed out in the image on the previous page (step #2). Once started, you will see the screen shown below. This practice exam software runs in your browser. Click the blue **Start Quiz** button to begin the practice exam.

Starting the practice exam

Next, you will enter your name and email address; this is optional. If an email address is entered, the practice exam results will be sent to that email address. If these fields are left blank, no results are sent. To skip this step, simply leave the data fields blank and click Submit.

No exam data is collected by the author or publisher of this book.

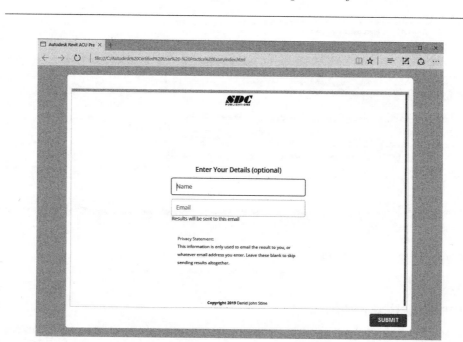

Enter your details (optional)

On the next slide, shown on the next page, be sure to note the information about entering your answer in the correct format. This is true for this practice exam and the actual exam at the test center. Your answer must match the format listed next to the box where you enter your answer.

The answer format is important

Keep in mind this can also help double-check your answer. For example, if you think the answer is **200 SF**, but the format indicated **###.## SF**, this means there must be a decimal value other than zero.

Once you click the **CONTINUE** button, shown below, the <u>timed</u> practice exam will begin. You will have **50 minutes** to answer 30 questions. If time runs out, the unanswered questions will be graded as incorrect.

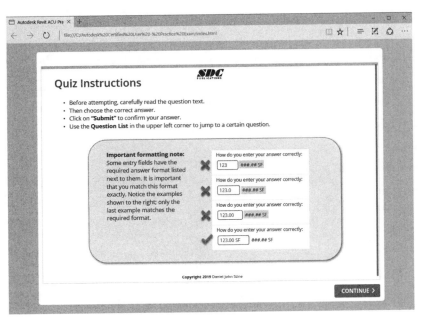

Exam instructions, format is important

One thing to keep in mind during the actual exam is you may need to use the Windows command **Alt + Tab** to toggle between the exam software, which takes up the entire screen, and Revit. To do this, simply hold down the **Alt** key, on the keyboard, and then tap the **Tab** key to cycle through the open Windows applications. Give it a try on your computer.

Practice Exam User Interface (UI)

The following image, and subsequent list, highlight the features of the practice exam's user interface.

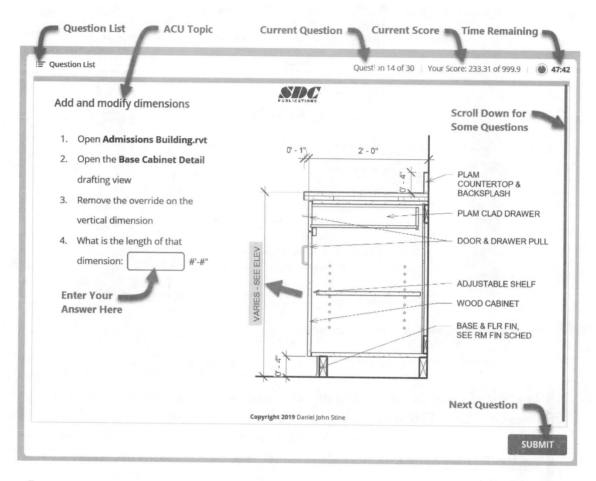

Practice exam highlights

User Interface details:

- **Question List:** Opens a menu listing all questions
- **ACU Topic:** Ties to exam topics and objectives list on page 5
- **Current Question:** Current number out of 30
- **Current Score:** Each correct answer is worth 33.33 points, 999.9 total points possible
- **Time Remaining:** Time remaining for the 50 minute timed exam
- **Enter Your Answer Here:** Type answer in this box
- **Scroll Down for Some Questions:** You must scroll down to see the full question
- **Next Questions:** Proceed to the next question

Practice Exam Results

When you complete the practice exam, you will find out if you passed or failed. The example image below indicates a failed attempt, including the overall percentage. Click the **Review Quiz** button to go back and see which questions you got wrong and study those topics again.

Practice exam results

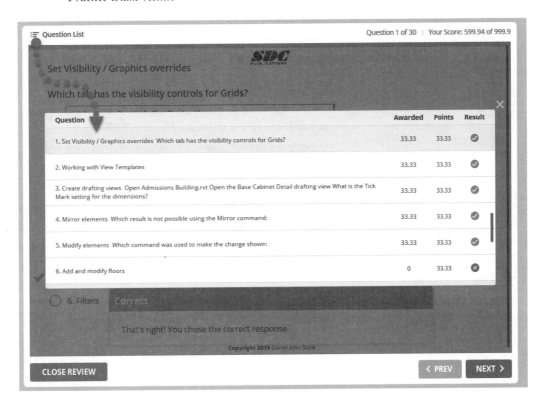

Practice exam results

While reviewing the results, a green check mark will appear next to the correct answer. Clicking the **Questions List** button in the upper left will provide an overview list of all questions as shown in the previous image. Note that the question name listed matches the topics and objectives covered in each chapter.

If you entered an email address, the practice exam results will be emailed. An example of this email is shown in the next image. Incorrect answers will list the page number to review in the book, making it easier to go back and study the relevant sections more.

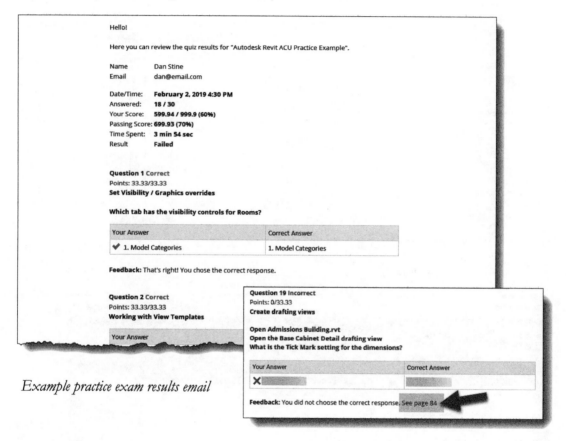

Example practice exam results email

Conclusion

As with any formal exam, the more you practice the more likely you are to have successful results. So, be sure to take the time to download the provided practice exam and give it a try before you head off to the testing facility and take the actual exam.

Good luck!

Index

The image below highlights the features of the practice exam software included with this book. See the inside-front cover for download instructions and your unique access code.

Question List   **ACU Topic**   **Current Question**   **Current Score**   **Time Remaining**

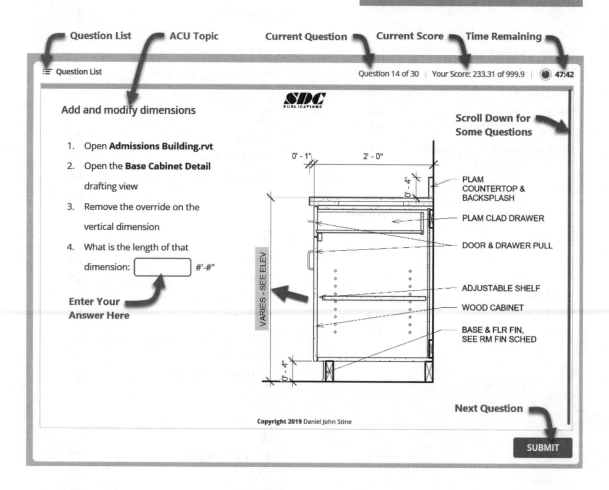

Question List   Question 14 of 30 | Your Score: 233.31 of 999.9 | ⏱ 47:42

SDC
PUBLICATIONS

Add and modify dimensions   Scroll Down for
                                                                              Some Questions

1. Open **Admissions Building.rvt**

2. Open the **Base Cabinet Detail**
   drafting view

3. Remove the override on the
   vertical dimension

4. What is the length of that
   dimension: [   ] #'-#"

**Enter Your
Answer Here**

0' - 1"   2' - 0"

PLAM
COUNTERTOP &
BACKSPLASH

PLAM CLAD DRAWER

DOOR & DRAWER PULL

VARIES - SEE ELEV

ADJUSTABLE SHELF

WOOD CABINET

BASE & FLR FIN,
SEE RM FIN SCHED

0' - 4"

Copyright 2019 Daniel John Stine

Next Question

SUBMIT